Dietmar P. F. Möller (Ed.)

System Analysis
of
Biological Processes

Advances in System Analysis
Editor: Dietmar P. F. Möller

Manuscripts submitted to *Advances in System Analysis* must be original, poiting out
the advancement of the contribution with respect to the actual a-priori knowledge.

Manuscripts or exposés should be sent to the Editor of the Series:
Dietmar P. F. Möller, Johannes Gutenberg Universität Mainz, Physiologisches Institut,
Saarstr. 21, D-6500 Mainz 1, W.-Germany.

Dietmar P. F. Möller (Ed.)

2nd Ebernburger Working Conference

Erwin-Riesch Workshop
System Analysis of Biological Processes

Bad Münster am Stein-Ebernburg

9.-11. April 1986

Springer Fachmedien Wiesbaden GmbH

Organisation:

Working Group 4.5.2.1 Simulation in Biology and Medicine of the Technical Committee
4.5 Simulation (ASIM) of the German Society of Informatics (GI) Dr. D. P. F. Möller
Physiologisches Institut, Johannes Gutenberg Universität Mainz, Saarstr.

Scientific Program Committee:

Prof. Dr. Dr. W. K. R. Barnikol, Universität Mainz
Prof. Dr. B. A. Gottwald, Universität Freiburg
Dr. D. P. F. Möller, Universität Mainz
Prof. Dr. O. Richter, Universität Bonn
Prof. Dr. R. Lunderstädt, Universität der Bundeswehr Hamburg

1987

All rights reserved
© Springer Fachmedien Wiesbaden 1987
Ursprünglich erschienen bei Friedr. Vieweg & Sohn Verlagsgesellschaft mbH, Braunschweig 1987

No part of this publication may be reproduced, stored in a retrieval system or
transmitted in any form or by any means, electronic, mechanical, photocopying,
recording or otherwise, without prior of permission of the copyright holder.

ISBN 978-3-528-08983-2   ISBN 978-3-663-19445-3 (eBook)
DOI 10.1007/978-3-663-19445-3

Preface

It is a great honour for me to open these Proceedings with a few brief remarks. Our 2nd Ebernburger Working Conference is devoted to the presentation and discussion of modern system analysis aspects of the entire field of biological processes.
The workshop was organized by the Working Group 4.5.2.1, Simulation in Biology and Medicine of the Technical Committee 4.5 Simulation (ASIM) of the German Society for Informatics (GI).
The name Erwin-Riesch Workshop deals with the generous support of this Workshop through the Erwin-Riesch Foundation. This support reduces the printing costs of the proceedings for the participants.

The papers included in this volume were presented at the old castle of Ebernburg in Bad Münster am Stein-Ebernburg at the 2nd Ebernburger Working Conference System Analysis of Biological Processes, from 9th to 11th of April, 1986. Without a doubt, the diversity of the program will appeal to the interdisciplinary research activities in the field of the analysis of bioprocesses. Therefore the proceedings are divided into four sections, the tutorial, the invited papers, the selected papers and the round table discussion.

Our meeting in Bad Münster am Stein-Ebernburg was one of the series of gatherings following earlier meetings in Bad Münster am Stein-Ebernburg in 1984 (organized by D. P. F. Möller) and Freiburg in 1985 (organized by B. A. Gottwald) and will be held in 1987 at Zürich (organized by S. S. Hacisalihzade), and 1988 again in Bad Münster am Stein-Ebernburg.

Bad Münster am Stein-Ebernburg with the castle of Ebernburg is a very famous small town at the river Nahe within the wine producing district of the well known Nahe wines. The castle of Ebernburg was destroyed twice, 1523 and 1698, and from 1794–1835 as quarry exploited. From 1954 to 1971 and 1974 to 1981 the castle was reconstructed in the old form of the second part of the 16th century.

The participants of the workshop living at the old castle of Ebernburg. The Workshop was opened wednesday evening by a wellcome receipion, given at the pleasant atmosphere of the Kurhaus of Bad Münster am Stein-Ebernburg under the sponsorship of the Kurdirektor Mr. Schneider. As pointend out briefly, Bad Münster am Stein-Ebernburg is a well known wine destrict, therefore the Kurdirektor handles to all participants and accompanying persons a bottle of Nahe wine for remembrance of this nice evening.

Thursday evening a knight banquet was given in the stimulating atmosphere of the old castle of Altenbaumburg. During the banquet, the international well known ensemble "Chantal", founded in 1968 at Alsey, play chamber-musically-rarities. This group was the winner of the price of the European Liberal Democrats in Brüssel, 1984, out of 94 groups from Europe.

The hight light of this evening were the concerto C-Dur of A. Vivaldi and a Bretonique melody arranged by M. Hofmann, the group leader, given at a candle light atmosphere. Many tanks to Chantal''.

Moreover we would like to thank Mr. Gattung and Mr. Ackermann of the Kurdirektion of Bad Münster am Stein-Ebernburg for their assistance, organizing the social program.

Also we would like to thank Mr. Rauschenplat and his co-workers for their excellent help to make the stay at the castle as pleasant as possible.

Finally our thanks are due to the speakers and sessions participants for stimulating presentations and discussions.

Last not least I would like to express my deepest gratitude to my wife and my daughter for their forbearance they gave to me during organisation this meeting and preparing the proceedings.

Mainz, Summer 1986   Dietmar P. F. Möller

Contents

Tutorial

Simulation of Biological Systems with Petri Nets – Introduction to Modelling of Distributed Systems

H. Fuss

Institute for Foundations of Information Technology (F1)
Gesellschaft für Mathematik und Datenverarbeitung (GMD) Bonn

Key Words and Summary

Notation: System, model and environment. Distributed real systems. The (maximal) signal propagation speed in a system. Simultaneousness: concurrency plus coincidence. The conflict situation, the glitch phenomenon. Different models, their aims: for planning and forecast; their types: material and abstract. Modelling vs. experiments. Analogy, correlation, timing, and causality. Time considerations in systems, models, and nets. An example of coordination of concurrent processes.

1   General Problems in Modelling Distributed Systems

In recent years an effect has become noticeable that is crucial for the functioning of some systems – and hence of equal importance to their modelling and simulation: a certain <u>disorder</u> – or more precisely: the partial order of causality.
This disorder is the other side of (partial) autonomy of the components of the system.

In some organisations, this partial autonomy is substantial for the systen (e. g. departments of a supermarket or of a bank: they can do business without letting the other department know), but it *could* be regulated by a central management (or registered by one central computer), if only the management so demanded.
But in some other organisations, as has become obvious in the last few years, it is technically and theoretically <u>impossible</u> to control centrally the functioning of the entire system, because the changes of the states at different places are too quick to keep track of, as even the highest possible signalling speed (which is the velocity of light) is limited. This has become obvious with the construction of electronic devices and astronautic space vehicles (e.g. after receiving a message that a meteorite had hit the space ship far beyond the moon or the sun, and sending signals from here in order to do some corrections at the space ship, some more events could have happened in space that make the intended actions obsolete).

In both cases, the essential point is that the interacting parts are spatially distributed and a real (not an idealized infinite) signal propagation speed is taken into account.

This paper contains parts of and an example from a publication of the third Sympos. on Automation and Scientific Instrumentation, held by the Bulgarian Academy of Sciences in Varna, Oct.'85

But it is not necessary to think of distances of billions of miles and of speeds like c to study this phenomenon, the same effect can be noticed in as tiny constructs as electronic chips or as popular organisations as an administration. We are already facing all the problems, once the *internal processing speed <u>within one</u>* component is <u>of the same order</u> as the *communication speed <u>between</u>* the components.

Under certain circumstances it could be useful to understand biological systems as belonging to this class of systems, for many biological phenomena in a body can happen in parallel in parts independent of each other. Especially with nerves, as with most information processing systems in general, the signal propagation speed <u>is</u> the maximal information processing speed.

Information processing systems usually have only <u>one</u> working speed, which is the maximal possible one in every component. This is different from a *goods processing* system: the production of (e.g.) cars and the physical transportation of car components is totally different from the signal 'transportation' speed in that factory.

In distributed systems, or more precisely: in *non-hierarchical* distributed systems, the aspect of <u>*concurrency*</u> (Nebenläufigkeit) plays a decisive role.

The coordination (synchronisation) of actions is <u>the</u> problem of distributed systems.

Very often the local separation is just *the* important constituent of the subsystem (e.g. stations in a railway system); sometimes it is only a very striking factor. Sometimes the separation is more on the organisational (informative) level (e.g. an administration). Very often both aspects apply (e.g. branches of a factory).

In all cases, the subsystems have different duties, functions, and interests which they are obliged to fulfil; and in order to do so, they have certain freedoms for their decisions. These decisions and their actions can conflict with those of other subsystems, they might even contradict each other. Furthermore, the speed with which messages are carried is finite, so there may be delays of unpredictable duration. It has been considered as a success of the *theory of distributed systems and concurrent processes*, (which is closely connected with the name of C. A. Petri) that these situations are recognized as system-immanent. Therefore, the methods of solving such problems must come from this theory.

Net theory is <u>one</u> model, probably the most advanced one, for dealing with concurrency. By now, several textbooks on net theory are available: [BRAMS], [Kotov], [Reisig], [Rosenstengel/Winand], [Silva], [Starke], [Zuse], etc., and several extensions or modifications on 'normal' Petri nets have been developed for some special purposes. Other approaches which have been persued to a great extent are: of algebraic type [Winkowski/Mazurkiewicz], or of relational type [Wedde], and path expressions.

2   Terminology: Systems, Models, and Processes

For clarity reasons, it is useful to remember the notion of $\mathcal{U}$, $\mathcal{R}$, $\mathcal{S}$, $\mathcal{M}$, $\mathcal{E}$, and $\mathcal{I}$.

Let $\mathcal{U}$ be the *universe*, and $\mathcal{R}$ be *reality*, that part of the universe which is known to us. We may think of $\mathcal{R}$ and $\mathcal{U}$ as being the same, for it does not make any difference here. But we should note that $\mathcal{R}$ is already a model of $\mathcal{U}$, every person might have a different $\mathcal{R}$ in mind.

One part of $\mathcal{R}$ is the *system S*, which is of special interest to us. The rest of S in $\mathcal{R}$ shall be called $\mathcal{E}$, the *environment* of the system, i.e. $\mathcal{E} = \mathcal{R} - S$. Communication between S and $\mathcal{E}$ will take place at certain points I, the *interfaces*. An interface may be of the type of a mailbox, where S or $\mathcal{E}$ deposit messages (or goods) for each other, or of the type of an interpreter, who translates from one language (or changes one currency) into the other.

When we want to do simulations, we first make a <u>model M</u> of the system, which is more precisely: a mapping of S to M. 'Unfortunately', M is part of $\mathcal{R}$ too, so we might get mixed up, but 'luckily' this is exactly the reason why we can iterate the procedure of modelling and have models of models.
Models may be as different as a model car, a set of differential equations on a sheet of paper, a computer program on a disk, some biological material in the brain...

Note: If S and M are the same (or nearly the same) part of $\mathcal{R}$, or if M is just a small part of S, then we would rather like to speak of *experimenting*.
Examples: chemists do so with little amounts of material (maybe in the environment of a simulation), physicists do the same, and sometimes physicians can't do anything but experiment on a patient.

If φ is the model-making mapping, i.e. if $M = \varphi(S)$, then (of course!) $\varphi(\mathcal{U}) \subseteq \mathcal{U}$, but: there is no shorthand notation for $\varphi(\mathcal{E})$ and none for $\varphi(I)$, there are not even separate words for it.

But mapping S to M is only one half of the job: we not only want to map the system precisely enough to a model, but we want to map the *behaviour of the system* to the *behaviour of the model* – and this is the far more difficult task.

A very successful way of achieving this seems to be to understand $\mathcal{R}$ and M as sets of **objects** (components, systems, subsystems), which are in certain <u>states</u>, and sets of **actions** which <u>change</u> these states (e.g. change numerical values). Such actions depend on the current states of the objects themselves which are the *causal* pre-conditions of the *events* (accordingly: post-conditions are the conditions after such an action).

The **structure** of both the system and its model is defined exactly by the way in which such objects and actors are combined.

Note: we are speaking of <u>causal</u> dependencies, not of an ordering of the events by time. A global *time* will not be assumed as an ordering relation. Even 'worse': the theory of concurrent events deals with *time* as a <u>partial</u> order, and it can even *explain* time as series of events which are produced by special physical instruments, commonly called oscillators or clocks.

It is rather easy to map the pure structure of the system to a model, even that of a distributed system. It is difficult to model the *processes* that run on distributed systems. But modelling the *behaviour* of such systems where human beings are involved (i.e. in socio-technical systems) is a very difficult task. As we cannot read people's minds, our choice of the alternatives might be different from theirs; i.e. in the models, the decision of conflicts (q.vis.) cannot be delegated to technical devices.
And in addition to the above, people, especially those who have to make important (economic, political...) decisions, will normally not disclose their future decisions, because they feel that this might impair their own chances, intentions, and interests.

3 The Basics of Petri Net Theory

3.1 Concurrency

To deal correctly with distributed systems – originals and models alike – we have to discuss *the* characteristic aspect of these systems: **concurrency**.

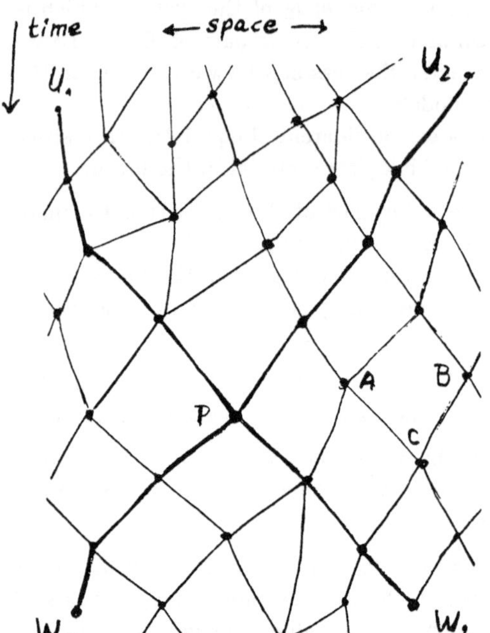

Concurrency $\underline{co} = (\neg prae) \cap (\neg post)$ describes the relation between events, when either event is **not before** *and* **not after** the other – both in a temporal, and more importantly in a *causal* way (i.e. they have no life-line in common). The description of the situation in elementary physics is well-known as the Minkovski cone. Let us consider point P. All events that can have an effect on P lie in the cone $U_1 - P - U_2$; all points where P might have an effect upon are in the cone $W_1 - P - W_2$. In particular, neither A nor B nor C are 'before' or 'after' P. That is, all three points A, B, and C are concurrent to P. But it would be a mistake to deduce herefrom that they are concurrent to each other. (It is $A \underline{co} B$, but (A, C) resp. (B, C) are on a *line* $\underline{li}$: $A \underline{li} C$, and $B \underline{li} C$.)

Beware of misunderstandings: Everyday language does not differentiate between the two relations 'concurrent' and 'coincident', they are both commonly named 'simultaneous'. But mathematically speaking, simultaneousness is an equivalence relation, but concurrency is a *similarity relation*, i.e. *concurrency is* <u>*not transitive*</u>.

Coincidence differs from concurrency in that the event happens 'at the same time <u>and</u> at the same location', it is at the intersection of two (or more) life-lines. Examples: clapping hands; lightning and thunder (which are created coincidently by *one* electric discharge but travel with different speeds). Counterexample: On jumping, a person will land concurrently on both his feet (it will vary depending on the ground, on his habits, etc., sometimes left, sometimes right foot first), maybe sometimes not distinguishable with the given observation tools and thus called 'at the same time', but definitely not 'at the same place' as well. (The landing pressure on one foot and the bump on the soil, however, will happen coincidently.)

3.2 Petri Nets in General (Place/Transition Nets)

It would be foolish to try to be as precise or as complete as a textbook on net theory. The idea here is to give a general overview of a few main ideas of net theory.

One important point is to see the difference between 'states' (Zustände) and 'changes' (Veränderungen), or, to be a little more precise, between the objects (which are in certain states), and the events that change the states – or even the objects.

The objects are usually graphically represented by circles $\bigcirc$, and the presence of the mentioned state is marked by a dot $\bullet$. Events are represented by boxes $\square$, with arcs $\longrightarrow$ connecting all objects which are involved in this specific event. At system level, where such events will occur repetitively, the boxes represent classes of events.

Obviously, circles and boxes are connected alternately, for there must be a change $\square$ between two different states, as well as a state $\bigcirc$ between two different steps $\square$.

The basic graphical pattern in net representations is the following one:

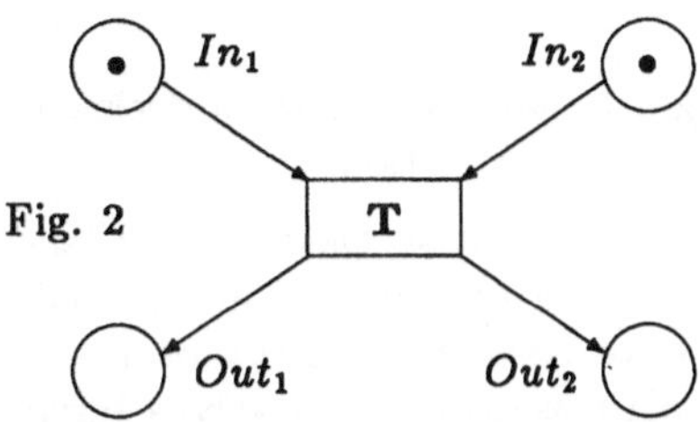

Fig. 2

The basic dynamic interpretation, according to the **firing rule** (Schaltregel) is as follows: If transition T fires (if all input places have been marked) – it clears all input places from and coincidently marks all output places with *tokens* $\bullet$, in the case of equal numbers of input/output paths, it looks like moving tokens (and like packing and unpacking otherwise).

The firing rule is called safe if no outputs are destroyed or overwritten.

3.3 Conflicts

The following diagram shows a typical situation which often occurs in the surroundings of a place A (here: for simplicity reasons restricted to the neighbours *after* A).

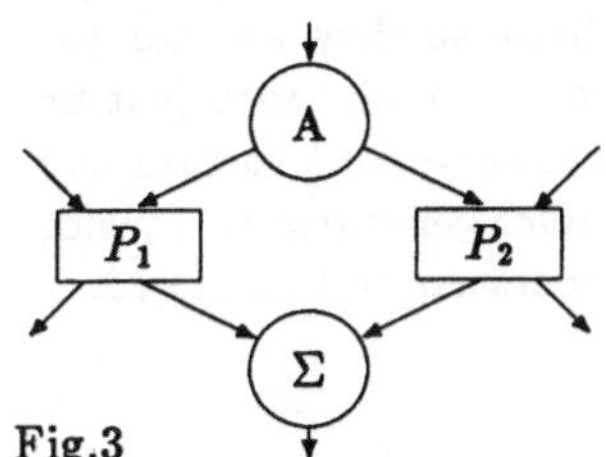

Fig.3

Obviously, the value of Σ depends on the path of the execution of the program. (Note: there is no *if*-statement!) P_1 and P_2 in general do different transformations on the value of A, depending on their own programs and on their further input. If we want to reverse the flow and to regain the value of A from a given Σ; then we must know the original path, and the influences on its way.

The above diagram can serve for the discussion of another problem, too:
Let us assume that P_1 and P_2 are two processes (persons) who simultaneously request for a resource which is available at A. Where should A go? Some more information than shown in the above diagram is needed for the decision. Semaphors or *if*-statements put this decision only one step ahead: how to set them?! For pending problems with priority rules see the glitch problem (below).

The **selection of one** out of two (or more) equally possible steps (alternatives) is called the **decision of a conflict.**

In several modelling languages, this information is generated from heuristic assumptions – which means: from nothing. Sometimes priorities are considered, sometimes a random number generator is involved. The reasons for deciding the conflict then are hidden or transferred to those who assign the priority or write the number generator program. So, the control over the result, i.e. the result itself is determined from outside the system.

If A is a person who himself has to decide which route to take he might run into the following, the glitch dilemma:

It is obviously a good policy for A to act after the rule 'first come first served' and serve P_1 or P_2 whoever is first. And in a case where both requests arise at the same time, A decides to toss a coin. He happily works with his rule till he has to throw his coin for the first time. Actually he did not get rid of the burden of deciding but in fact doubled it: he covered the interval of uncertainty with a seemingly safe algorithm, but on both ends of that algorithm there are new unclear borders between the possibilities to serve FIFO, or to toss a coin.

This shows that any technical device, though it might be able to suggest a decision, cannot relief A from carrying the *responsibility* for his (or her) action.

The notion of conflict is unknown to most of the other simulation methods and languages. This is the most striking difference between nets and more 'conventional' methods. In nets, **all conceivable courses of events have been provided for**, though they may never occur; in the other methods **all possible alternatives have been decided**, though in reality some of the decisions may take place in a different (or unknown) way.

The most common way of dealing with uncertainty about the decision of conflicts, or about the knowledge thereof, is to use statistics and to weigh the different paths through the program by probabilities. But to do so is only satisfactory if the 'law of large numbers' applies. Sometimes, e.g. for planning purposes [Fu3], it might be desirable to discuss one single path through the model and to reflect on every single point of decision. In those cases statistics are of no help, because they are not yet available. And considering carefully every single step of the development may just be the task to be solved: planning the system and its functioning to serve for a certain real problem. For such purposes, the application of a random number generator will yield, as a result of the simulation, just random numbers, nicely processed and arranged.

3.4 Special Forms of Petri Nets

After their 'invention', Petri nets as a mathematical structure themselves became an object of mathematical-theoretical studies. Different practical problems demanded different interpretations, and different common properties were found. A first general classification was suggested and discussed[1] in order to find a common terminology.

What we discussed here and what seems useful for simulation purposes in a biological context are **place/transition-nets**. One characteristic feature is that the places are of (finite) capacity, in general it is $c(P) > 1$.

More useful for simulations of technical systems seem to be condition-event-nets, or, according to the notation mentioned above, **condition/event-systems**. In those

---

[1]Best/Fernandez on the 6th Europ. Workshop on Applicat. & Theory of Petri Nets, Helsinki, June '85

nets, the capacity of places is 1 token, and a multiplicity of tokens (signals) is transposed into the arcs.

In order to describe a structure in brief, **channel-office-nets** have been useful.

Well known from automata theory are **state machines** which are nets with no branches of control (flow) in the boxes (transitions). The dual type of nets with no branches in the places is named **synchronisation graphs**. Some other classes of nets with common properties are free choice nets, occurrence nets, where some authors speak of causal nets; rather popular for describing intricate structures are **predicate/transition-nets**, etc.

One property of nets which is of some importance for research is the absence of *side conditions*, such nets are named pure nets.

3.5 Place-Transactor-Nets or P/Ta-Net

One type of nets which is worthwhile mentioning here because they can well be used for numerical simulation are Place-Transactor-Nets. The aim is to have a variabe token width, i.e. a parametrized 'intensity' of the flow $\mathcal{F}$ between the Places. This is achieved by the function of some Places as parameter-containing K-Places.

The Statics of a P/Ta-Net

A K-Places, which contains the values of parameter, is adjoined to every arc.

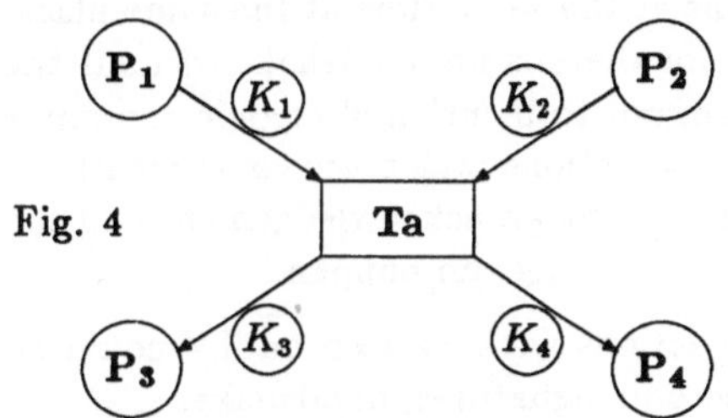

Fig. 4

The Place itself is always the same object, it is only its *pragmatic status* which varies with its position in the flow relation in that case, i.e. with its use (similar to a noun in a simple sentence: it may be an object (P), or a subject (K)).

The value of the parameter, i.e. the contents of a K-Place can be changed, once it is used ('switched') as an ordinary P-Places.

The Dynamics of a P/Ta-Net – The Dynamisation of the Token Flow

The token flow between P-Places is ruled by the contents of the associated K-Places at the 'time' of the transaction (i.e. its occurrence) in a one-to-one way: the number of tokens moved to or from a P-Place by a transaction is equal to the number of tokens which are the contents of the associate K-Place *at that time*.

'During' the transaction, the value of a K-Place remains unchanged, which is important in the case of self-relation.

Let..

$c(P_n)$ be the capacity of a Place P_n, $c(P_n)\epsilon\ \mathcal{N}_1$, with $\mathcal{N}_1 = \{1, 2, 3...\}$,

$i(P_n)$ be the contents (number of tokens which are in P_n), with $0 \leq i(P_n) \leq c(P_n)$,

$\triangle(P_n)$ be the defect (difference) $\triangle(P_n) = c(P_n) - i(P_n)$;

then for any Transactor Ta_0 the Transaction Rule R is :

<u>If</u> $\sum_{j} i(K_{1,j}) = \sum_{j} i(K_{2,j})$ for all $K_{1,j}$ and $K_{2,j}$ which are in F_1 and F_2 of Ta_0
<u>and</u> $0 < i(K_{1,j}) \leq i(P_{1,j})$ for every input pair $(P, K) \subset F_1$
<u>and</u> $0 < i(K_{2,j}) \leq \triangle(P_{2,j})$ for every output pair $(K, P) \subset F_2$
then the Transactor Ta_0 is <u>*enabled*</u>, and *if* it fires, the result of the transaction is:

$$i(P_{1,j}) - i(K_{1,j}) \Longrightarrow i(P_{1,j})$$ for every in-pair $(P, K) \subset F_1$ <u>and</u>
$$i(P_{2,j}) + i(K_{2,j}) \Longrightarrow i(P_{2,j})$$ for every out-pair $(K, P) \subset F_2$.

4   Application of a Petri Net Model: The Signature of a Treaty by Two State Officials

The situation is quite familiar, because we know it from television: the two high-ranked persons P1 and P2 (Prime Ministers, Secretary of State or of the Central Committee, etc.) sit at the table, functionaries F1 and F2 display engrossments of a document to them, they sign it, F1 and F2 dry the ink of the respective signature, then pass the document to the other officials (P2 and P1 resp.) who now sign it, (again drying of ink,) finally the politicians shake hands.
(Legal question: When does the treaty get the status 'signed'?)

For simplicity reasons, we concentrate on the politicians and the documents only.

In former times, when legal questions and arguments of lawyers were of less importance than nowadays, the handshake was the symbol of a deal – and this for a very good reason: this event is a coincident one, it happens at the same time at the same place: both partners have to be present and to agree, for there is no handshake without the other partner. But nowadays in times of written, not to mention electronic computer communication, acknowledgements of relevant interactions take place concurrently at different places at different times. And between sending an acknowledgement and the time his partner gets the knowledge of it — many a thing can happen.

A very simple system description model of the actions we have seen on TV could be the following sequence: signatures, exchange, second signatures, handshake.

This is really a very crude one, and it describes only very inaccurately what really happens, for the first two actions and the following two happen somewhat 'in parallel' and 'overlapping' but neither starting nor ending 'at the same time', the overlapping is at the discretion of each of the two high-ranking persons.

To prevent possible misunderstandings, we clearly want to point out the following:

a) if we register the <u>history</u> of these events, then we can (within the precision of our measuring instruments) tell the amount of overlapping, i.e. tell the temporal sequence (though it depends on the observer)

b) if we want to make a model for the future occurrences of such structured events we have to leave open all restrictions on the sequences of these actions.

Below we give a description of the above processes in the language of a Petri net which is suited to describe correctly concurrent processes:

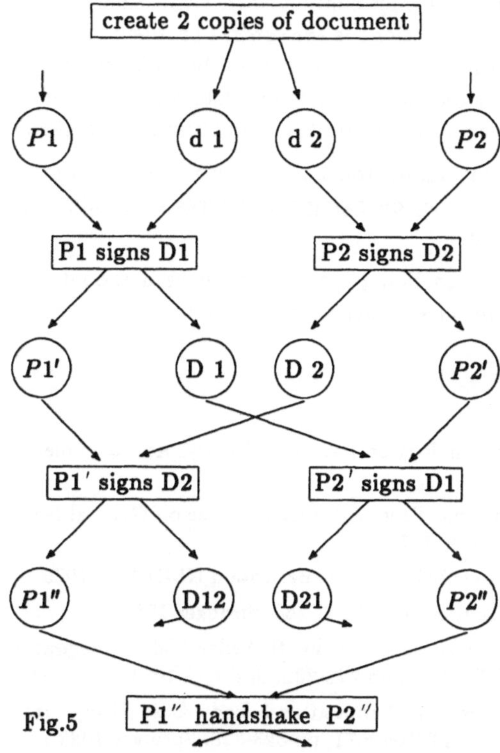

Fig.5

The empty forms d1 and d2 obtain different pragmatic status after one signature (D1), (D2); and after two signatures (D12), (D21), and so do the state officials: they would not sign the same engrossment a second time. Sometimes it seems peculiar why 'the same' objects under different circumstances are represented by different circles (or 'the same' actions by different boxes). The explanation is: they are 'the same' yet seen at different times. But the state of an object whether before or after an action is sometimes so important that even natural languages have different names for it: before a person buys something he is a prospective customer, after purchase he is the owner of it; in chess it is the threatening and the move, here: a draft and a document, etc. Both politicians are free to sign independently from their partner. To insist on their freedom of action, and to model it, is of demonstrational value for us. They really run concurrent processes.

Their behaviour, even if they sit within sight of each other at the same table, can be seen as a model for the situation where two persons operate at two remote terminals on the same data base of a computer. Hardly any synchronisation is then possible by social agreement, because usually the terminals are not within sight of each other.

5 CONCLUSION

No model can produce more information than it knows itself, – it can only process and transform information. Therefore it does not make much sense to speak of sources and sinks of information other than of entry and exit points of information (i.e. of interfaces) between system and environment – and this applies to a simulation system as well. The corresponding inverse argument is: if one leaves out a set of original system information when he builds a model, then the simulation system lacks this aspect and is likely to produce poorer results.

In some real systems the fact that they are *distributed systems* might be a constituent of the system. For better simulation results, modelling should be done without destroying this structure. Unfortunately, the problem scope in these systems is extended by a class which is unknown to (clocked) sequential systems. As we have seen, the problems in distributed systems derive from the fact that the modules (i.e. the subsystems) interact with each other, and these interactions may conflict with each other. The coordination and synchronisation of such (inter)actions is *the* central problem of their processes. By using inappropriate models the problems are 'solved' by definition: by defining them out of the model. Net theory is one of the most successful ways of giving correct descriptions and models of distributed systems and their concurrent processes.

Despite the fact that there are not yet that many tools available as in the traditional modelling and systems theory, some very encouraging experiences in simulating distributed systems have been made with nets.

And in addition to the above: we regard it as being essential to base a model on mappings of **causal structures** rather than on similarities or correlations.

6 REFERENCES

[Be/Fe] E.Best,C.Fernandez: Notations and Terminology on Petri Net Theory. Arbeitspapiere der GMD No.195, 29 S. GMD (1986)

[Br] W.Brauer (Ed.): Net Theory and Applications. Proc. Advanced Course on General Net Theory of Processes & Systems. LNCS 84; Springer 1979

[Fu1] H.Fuss: AFMG - Ein asynchron. Fluss-Modell-Generator. Berichte d.GMD 100. 1975

[Fu2] H.Fuss: Verteilte Simulationen. in: Inf.FachBer.56, pp.283-288, Springer 1982

[Fu3] H.Fuss: Reversal Simulation with Place-Transactor-Nets. in: H.Wedde (Ed.): Adequate Modeling of Systems. Proc.Int.W.Conf. Model Realism, pp.222-232, Springer 1983

[Fu4] H.Fuss: Simulation of Distributed Systems - A Competitive 3-Body Case Study. in: W.Ameling (Ed.): Proc. 1.Europ. Sim.Congr.'83. Inf.FaBer.71, pp.323-328; Springer 1983

[Fu5] H.Fuss: Improvement of Simulation by Place/Transactor-Nets. in: A.Jávor (Ed.): Simulation in R&D, pp.85-91. IMACS. N.Holland Publ.Co. 1985

[Fu6] H.Fuss: General Aspects in Modelling Distributed Systems. in: D.Burev (Ed.): 3.Int. Symp. Automation and Scientific Instrumentation'85, pp.38-49. CLANP, BAN, Sofia 1985

[MS&] A.Maggiolo-Schettini/H.Wedde/J.Winkowski: Modeling a Solution for a Control Problem in Distributed Systems by Restrictions. TCS 13, pp.61-83, N.Holl. 1981

[Pe1] C.A.Petri: Concurrency as a Basis of System Thinking. in: Jensen/Mayoh/Moller (Eds.): Proc. 5. Scandinav. Logic Symp., pp.143-162. Ålborg Universitetsforlag (1979)

[Pe2] C.A.Petri: State-Transition Structures in Physics and Computation. Int.J. Th.Physics, Vol.21, No.12, pp.979-992

[PSI] NET - a Petri Net Editor and Simulator. Technical Report d. Fa. PSI, Berlin. 1985

[Re] W.Reisig: Petrinetze. Eine Einführung. 158 S. Springer 1982; also:
Petri Nets (*in English*). 150p. EATS Monogr. on Th.Comp.Sc. Vol.4. Springer 1985

[Re4] W.Reisig: Systementwurf mit Netzen. 125 S. Springer, 1985

[We1] H.Wedde: Lose Kopplung von System-Komponenten. Ber. d.GMD 96; GMD (1975)

[Wi] J.Winkowski: Algebras of Partial Sequences: A Tool to Deal with Concurrency. in: M.Karpinski (Ed): Fundamentals of Computing Theory. Springer, New York 1977

[P/P] E.Pless/H.Plünnecke: A Bibliography of Net Theory. Internal Report ISF-80.05, 2nd Edition. GMD (1980) (a new edition is in preparation for 1986)

[*NLs*] (quarterly) *Newsletters* of the SIG Petri Nets and Related System Models, GI – Gesellschaft f. Informatik, Bonn (ISSN 0173-7473)

Invited Papers

Synergetics and its Application to Biological Systems

H. Haken and A. Wunderlin
Institut für Theoretische Physik
der Universität Stuttgart

Abstract:

We shall discuss the mathematical formulation of the slaving principle and present a biological application: The theory of phase transitions in human hand movements.

1. Introduction

One of the most fascinating aspects of synergetics is provided by its interdisciplinary aims. Indeed systems from quite different disciplines have been discussed sucessfully by using the unifying viewpoint which has been introduced through the formulation of the slaving principle [1], [2]. The principle states that in the vicinity of a critical region complex systems are govered by only few collective degrees of freedom, the so-called order parameters, which dominate the macroscopic behaviour of the system. As an immediate consequence one may conclude that distinct systems which are composed of subsystems of quite different nature will behave in an analogous fashion. This has been confirmed by a detailed analysis of various systems from natural, technical, and social sciences. Here we shall only mention some of them. In physics the most prominent example is the laser process [3] which can be understood as a spontaneous selforganization of laser-active atoms through their interaction with light. An example from hydrodynamics is provided by the convection instability in the Benard problem [1], [2]. This problem has various applications especially in meteorology and the physics of the earth. To give an example from chemistry we refer to the Belousov-Zhabotinskii reaction [4] and in biology special attention has been payed to models in morphogenesis [1], the human brain [5] etc.

This article is organized as follows. Section 2 gives an elementary outline of laser theory where the ideas of synergetics can be discussed from first principles of physics up to the macroscopic process of collective ordering which is connected with the laser action. Section 3 formulates the general methodology of synergetics and gives special attention to the slaving principle. This will be applied in Section 4 to the problem of human hand movement.

The laser may be considered as a paradigm of synergetics. Its theory has been developed from first principles and experimentally verified in great

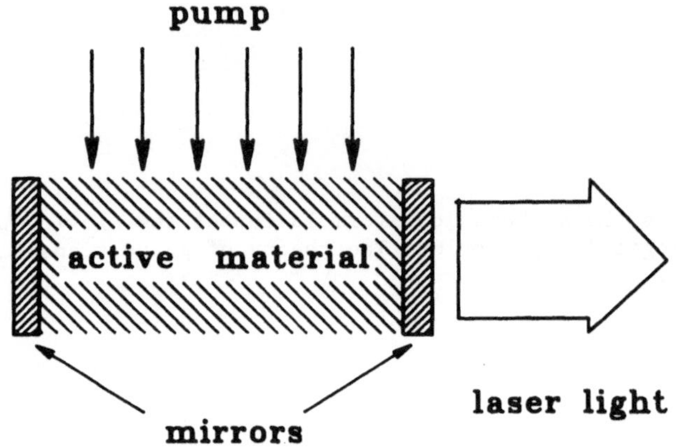

Fig.1: Model of a solid state laser

detail [3]. As a model we take the solid state laser (Fig. 1). In between two mirrors, acting as a cavity, there is a host crystal in which the laser active atoms are embedded. They are assumed to be separated far enough from each other and their mutual wave functions do not overlap. They therefore can be considered as independent. For the sake of simplicity we shall treat them as two level atoms. They interact with the

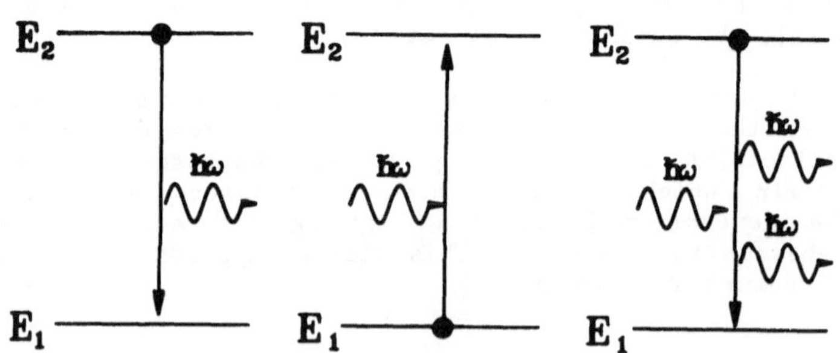

Fig.2: Interaction of light and matter

light through the following processes (Fig. 2): the spontaneous emission(a), absorption (b), and induced emission (c).

Following the notions of synergetics we shall consider the laser as an open system: There is a flux of energy through the system which is due to the pumping process. Furthermore we note that the system is composed of many subsystems, namely the laser-active atoms. We now shall convince ourselves that the laser process is connected with a spontaneous ordering of the laser active atoms over the whole device.

In order to introduce a complete set of variables to describe the system mathematically we first shall consider the light field. This may be described by the corresponding electric field $\mathbf{E}$ which is a function of position $\mathbf{x}$ and time t. It will be decomposed into cavity modes $\mathbf{u}_\lambda(\mathbf{x})$

$$\mathbf{E}(\mathbf{x},t) \;=\; \sum_\lambda N_\lambda (b_\lambda(t) + b_\lambda^*(t)) \mathbf{u}_\lambda(\mathbf{x}), \tag{2.1}$$

where N_λ is a normalization constant and $b_\lambda(t)$ are the amplitudes of the modes λ. We shall confine ourselves to the semiclassical approach where the field is treated classically.

The atoms, enumerated by the index μ, are described by their individual polarization $p_\mu(t)$

$$p_\mu(t) \;=\; - \alpha_\mu(t)\Theta_{12} \;-\; \text{c.c.} \tag{2.2}$$

Θ_{12} denotes the dipole matrix element between the states 1 and 2 and the time dependent amplitude. Finally we have to take into account the inversion of each atom which measures the difference in population N_i (i = 1, 2) of the upper and the lower level of the μ-th atom:

$$d_\mu(t) \;=\; (N_2 - N_1)_\mu. \tag{2.3}$$

Combining Maxwell's theory of the electromagnetic field with the quantum mechanical theory of the dipole interaction between light and matter we obtain the following set of equations

$$\dot{b}_\lambda \;=\; - (i\omega_\lambda + \kappa_\lambda)b_\lambda \;-\; i \sum_\mu g_{\mu\lambda}^* \alpha_\mu \;+\; F_\lambda, \tag{2.4}$$

$$\dot{\alpha}_\mu \;=\; - (i\nu + \gamma)\alpha_\mu \;+\; i \sum_\lambda g_{\mu\lambda} b_\lambda d_\mu \;+\; \Gamma_\mu, \tag{2.5}$$

$$\dot{d}_\mu \;=\; \gamma_\|(d_o - d_\mu) \;+\; 2i \sum_\lambda (g_{\mu\lambda}^* \alpha_\mu b_\lambda^* - \text{c.c.}) \;+\; \Gamma_{d_\mu}. \tag{2.6}$$

In these equations damping is included: κ_λ measures the losses of the field mode, γ is determined from the atomic live time, $\gamma_\|$ is the damping constant of the inversion. The field modes oscillate with frequencies ω_λ and the dipole moments with frequency ν. The coherent part of the interaction shows the forcing of the light field via the sum of the atomic dipole moments in (2.4) and $g_{\mu\lambda}$ is the corresponding coupling constant. The generated field acts via the dipole moments on the inversion (2.6) which then together with the field acts back onto the dipole moments again (2.5). Inevitably connected with dissipation are fluctuations which are taken into account by the fluctuating forces F_λ, Γ_μ and Γ_{ϕ_μ}.

The equations furthermore contain an external parameter d_o which measures the strength of the pumping rate which may be controlled from

outside. The whole set of equations may be interpreted as a set of
nonlinear stochastic differential equations.

Before solving these equations approximately we shall roughly discuss the
experimental observation which has been predicted theoretically [3]. In
the case of small d_o the laser is only slightly driven from thermal
equilibrium. Dominating is the process of spontaneous emission which is
purly stochastic in its nature. The electric field E(t) has a shape as it
is schematically drawn in Fig.3, the laser is acting as a usual lamp. By

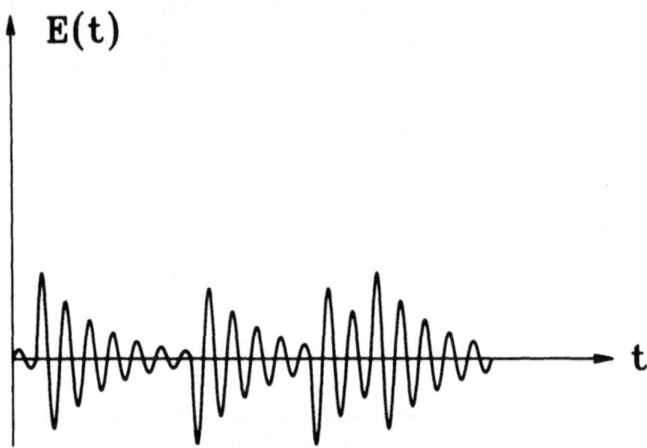

Fig.3: Electric field as a function of time below theshold

raising d_o, however, we pass a critical value $d_o = d_c$. Beyond d_c
laser action occurs. The observed field then is illustrated in Fig.4: One
observes a large wave track with a definite frequency and a highly

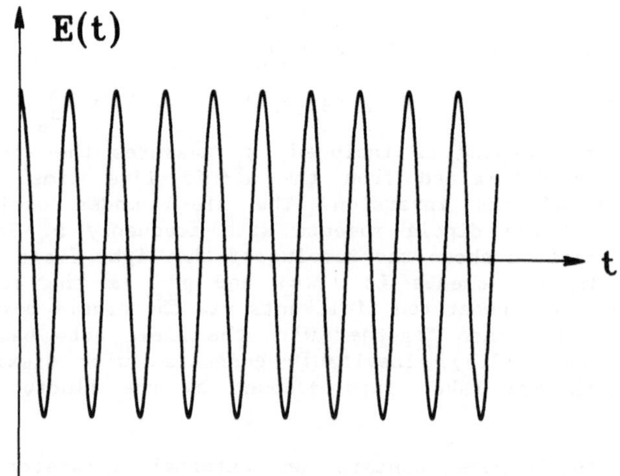

Fig.4: Electric field above threshold as function of time

stabilized amplitude. This effect obviously can be understood in the following way: There is a spontaneous ordering of the laser- active atoms on a macroscopic scale. All atoms are acting in a well defined way to produce the coherent wave. Clearly these atoms need "information" how they have to behave in a correct fashion. Obviously it must be the generated light wave which introduces the information about the correct phase behaviour of the subsytems.

We shall try to explain these observations from the basic equations (2.4 - 2.6). Here we shall confine ourselves to a special case, where the inversion is homogeneous

$$d_\mu = d. \tag{2.7}$$

It is then possible to prove that only one mode λ can be macroscopically excited. Preferred is the mode which is in resonance with the atomic transition frequency. Taking this for granted we may supress the index λ. We now can introduce collective variables in a straight forward manner. We use

$$S = \sum_\mu g_\mu \alpha_\mu \tag{2.8}$$

and furthermore

$$D = \sum_\mu d_\mu = \sum_\mu d = Nd; \qquad D_o = Nd_o, \tag{2.9}$$

where N denotes the number of laser active atoms. Transforming the variables to a rotating frame

$$S = \tilde{S} e^{-i\nu t}, \quad b = \tilde{b} e^{-i\nu t} \tag{2.10}$$

and dropping the tilde we obtain

$$\dot{b} = - \kappa b - iS + F, \tag{2.11}$$

$$\dot{S} = - \gamma S + i|g|^2 Db + \Gamma, \tag{2.12}$$

$$\dot{D} = \gamma_\parallel (D_o - D) + 2i(Sb^* - c.c) + \Gamma_D. \tag{2.13}$$

For the moment beeing we shall neglect the fluctuations and concentrate to the systematic motion. To that end we note the existence of the following hierarchie in time scales

$$\kappa \ll \gamma_\parallel \ll \gamma. \tag{2.14}$$

The slaving principle now states that the slowest variable will dominate the behaviour of the complex system. We may use this fact to solve the equations for the fast variables approximately.

We integrate (2.12)

$$S = \int_{-\infty}^{t} \exp(-\gamma(t-\tau))i|g|^2 D(\tau)b(\tau)d\tau \tag{2.15}$$

(long term behaviour) and make use of the hierarchy (2.14)

$$S = D(t)b(t) \int_{-\infty}^{t} i|g|^2 \exp(-\gamma(t-\tau))d\tau + O(\frac{1}{\gamma^2}). \tag{2.16}$$

Performing the elementary integration we obtain

$$S = i|g|^2 D(t)b(t) \cdot \frac{1}{\gamma}. \qquad (2.17)$$

We now notice ideed that the result is a consequence of slaving: The long living variables dominate the behaviour of the short living ones. The set of differential equations (2.11 - 13) may now be reduced to

$$\dot{b} = -\kappa b + \frac{|g|^2}{\gamma}D(t)b(t) + O(\frac{1}{\gamma^2}). \qquad (2.18)$$

and

$$\dot{D} = \gamma_{\parallel}(D_0 - D) + 2i\left(\frac{i|g|^2}{\gamma}D(t)|b(t)|^2 - c.c\right) + O(\frac{1}{\gamma^2}) \qquad (2.19)$$

In a similar way we may apply the idea of slaving again by taking into account the first part of the hierarchy (2.14). Integration of (2.19) yields

$$D = D_0 - \frac{4|g|^2}{\gamma}\int_{-\infty}^{t}\exp(-\gamma_{\parallel}(t-\tau))D(\tau)|b(\tau)|^2 d\tau. \qquad (2.20)$$

This equation may be integrated by the standard procedure to solve Volterra integral equations iteratively:

$$D \approx D^{(1)} + D^{(2)} + \ldots \qquad (2.21)$$

Then

$$D^{(1)} = D_0 \qquad (2.22)$$

and

$$D^{(2)} = \frac{4|g|^2}{\gamma}\int_{-\infty}^{t}e^{-\gamma_{\parallel}(t-\tau)}D_0|b(\tau)|^2 d\tau \qquad (2.23)$$

$$= \frac{4|g|^2}{\gamma\gamma_{\parallel}}D_0|b(t)|^2 + O(\frac{1}{\gamma_{\parallel}^2}).$$

Putting this result into the equation for the field mode (2.18) we are left with

$$\dot{b} = \left(-\kappa + \frac{|g|^2}{\gamma}D_0\right)b - \frac{4|g|^4}{\gamma^2\gamma_{\parallel}}|b|^2 b. \qquad (2.24)$$

Equation (2.25) represents the basic result which we obtained from the application of the slaving principle. The complete time dependent action of the complex laser system can be understood from one equation of motion. b, the amplitude of the surviving mode, takes the role of the order parameter. All subsystems, the single atoms, are dominated in their action by the order parameter b.

Obviously we may write

$$\dot{b} = -\frac{\partial V}{\partial b*} \qquad (2.25)$$

and up to an arbitrary constant

$$V = \left(\kappa - \frac{|g|^2}{\gamma}D_0\right)|b|^2 + \frac{2|g|^4}{\gamma^2\gamma_{\parallel}}|b|^4. \qquad (2.26)$$

For the sake of simplicity we now assume b real. (2.25) may then be interpreted as the overdamped motion of a particle with coordinate b in the potential V (Fig.5). The minima of the potential represent the

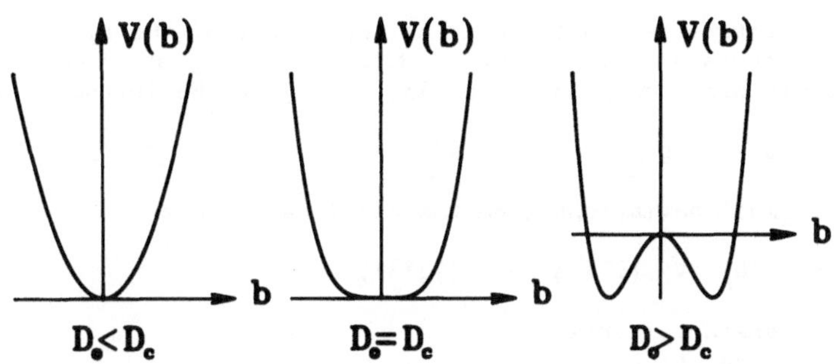

Fig.5: Potential V for different values of the pumping
       parameter D_o

stationary stable states. By changing the external parameter D_o we continuously follow up a sequence of potentials, three typical of them are exhibited in Fig.5. The transition may be identified with a structural instability in the strict mathematical sense. The qualitatively different states (in the sense of topological equivalence) correspond to the action of the laser as a usual lamp and the collectively ordered state of laser action, respectively.

3. The general Method of Synergetics

Synergetics deals with systems which are composed of many subsystems. These systems have the ability of a spontaneous selforganization on a macroscopic scale. A systematic treatment starts from the behaviour of the subsystems and their mutual interactions. On a mesoscopic level we expect "classical" equations of the general form

$$\dot{U} = G(U, 'V', \{\sigma\}) + F. \tag{3.1}$$

U is a vector of a state space Γ which is spanned by all of the degrees of freedom of the complex system. G is a nonlinear function of the state vector and in addition may depend on spatial gradients of U which is symbolically indicated through $'V'$. Finally G will depend on a set of external parameters $\{\sigma\}$ which are controlled from outside. F indicates fluctuating forces.

Here we shall confine ourselves to the most simple situation where the dimension of the state space is large but finite and where there are no spatial gradients.

Usually a complete global analysis of the Equations (3.1) is impossible.
We therefore concentrate us to special regions in the state space, the
so-called attracting sets. They are identified with the stationary states
U_o of the system under consideration. In general they may be quite
different in nature. In the most simple cases U_o is homogeneous in
space and time. The solution may then be identified with the thermodynamic
branch corresponding to the system. In more general situations, however,
it can be space dependent, periodic in time, or quasiperiodic.

We are mainly interested in the critical behaviour of the systems and thus
are led to discuss the stability of the underlying stationary state U_o
which we assume here homogeneous in time and space. We introduce q via

$$U = U_o + q \qquad (3.2)$$

and consider small deviations from the stationary state

$$\dot{q} = L(U_o,'\nabla',\{\sigma\}) \, q + O(\|q\|^2), \qquad (3.3)$$

where L is a constant matrix

$$L_{ik} = \left. \frac{\partial G_i}{\partial U_k} \right|_{U=U_o} \qquad (3.4)$$

which still depends on the control parameters $\{\sigma\}$. Through the ansatz

$$q = q_o \exp(\lambda t) \qquad (3.5)$$

(3.3) in our case can be reduced to a purely algebraic problem. The state
U_o is stable as long as

$$\mathrm{Re}\ \lambda_i < 0, \qquad \forall\ i. \qquad (3.6)$$

Instability occurs in the vicinity of critical regions in parameter space
which are determined from

$$\mathrm{Re}\ \lambda_i(\{\sigma\}) = 0. \qquad (3.7)$$

Linear stability analysis yields the critical regions corresponding to
U_o in the parameter space. However, we still may gain more
information. To each eigenvalue λ_i there corresponds an eigenvector
O_i and the set of eigenvectors is complete. These eigenvectors are
identified with the collective modes of the system. We therefore may apply
the hypothesis

$$q = \sum_i \xi_i(t) \, O_i \qquad (3.8)$$

with up to now undetermined time-dependent amplitudes $\xi_i(t)$. From the
linear analysis the amplitudes ξ can be classified in the following way.
In the vicinity of critical regions we find eigenvalues λ_u which
become unstable and others λ_s with still negative real parts. Obviously
there now exists a hierarchy in time scales

$$\frac{1}{\mathrm{Re}\ |\lambda_s|} \ll \frac{1}{\mathrm{Re}\ |\lambda_u|} \qquad (3.9)$$

and correspondingly we may split

$$\{\xi\} \begin{array}{c} \nearrow \text{ u} \\ \searrow \text{ s} \end{array} \qquad\qquad (3.10)$$

where generally there a few amplitudes u and many amplitudes s. Now inserting (3.8) into the full nonlinear equation for q and using (3.10) we obtain two distinguished sets of equations

$$\dot{u} = \Lambda_u u + A_{uuu}:u:u + 2 A_{usu}:u:s + \ldots , \qquad (3.11)$$

$$\dot{s} = \Lambda_s s + A_{suu}:u:u + \ldots , \qquad (3.12)$$

where the u's are varying on a much larger time scale than the s. According to the slaving principle the short living modes s are dominated by the long living modes u which means

$$s = s(u) + s_1 \qquad\qquad (3.13)$$

i.e., the stable modes are completely determined by the instantaneous values of the long living modes u. The term s_1 has been added to take additionally care of the fluctuations around $s(u)$.

There are standard procedures to determine $s(u)$ as well as s_1 [1], [6], [7]. Taking this for granted we immediately can obtain the order parameter equations by inserting (3.13) into (3.11):

$$\dot{u} = \Lambda_u u + A_{uuu}:u:u + 2 A_{usu}:u:(s(u) + s_1) + \ldots . \qquad (3.14)$$

The emergence of new ordered states is now connected with new nontrivial solutions of the order parameter equations. We furthermore observe that the slaving principle allows a drastic reduction of the original set of equations to few equations for the collective modes which dominate the behaviour of the whole system.

We note that the concept of slaving goes far beyond the purely mathematical theories of bifurcation and center manifold. Conceptually it is a principle of nature yielding the reduction to few collective degrees of freedom and includes time dependence as well as fluctuations.

In cases where the detailed nature and couplings of the subsystems are not known in detail, as it especially may happen in biological systems, the slaving principle still remains a powerful tool. Indeed, near critical regions the detailed nature of the subsystems becomes unimportant. We are therefore allowed to derive order parameter equations by purely phenomenological reasoning. This will be done for a concrete biological model in the next section.

4. Modelling of Complex Systems by means of the Order Parameter and Slaving Concepts, an Example from Biology

When we observe the motion of humans or animals, we are struck by their precise movement which is based on an extremely high coordination between the activities of neurons and muscles.

One may speculate that this high coordination is produced by specific programs, socalled motorprograms, of the neuronal net. More recent *experiments and their interpretation* by means of synergetics indicate rather different mechanisms, however. To this end let us consider a rather simple, though still quite surprising, experiment which was done by Kelso [8].

He asked test persons to move their fingers in parallel and then he asked them to move their fingers more quickly (Fig.6). Surprisingly it turned

Fig.6: Fast, in phase movement of two fingers

out that when the frequency of the finger oscillation was increased a sudden involuntary change of the mode of motion occurred.

Namely, instead of a motion in the form as shown in Fig.7, a new symmetric motion occurred (Fig.6). Let us try to model this abrupt change of

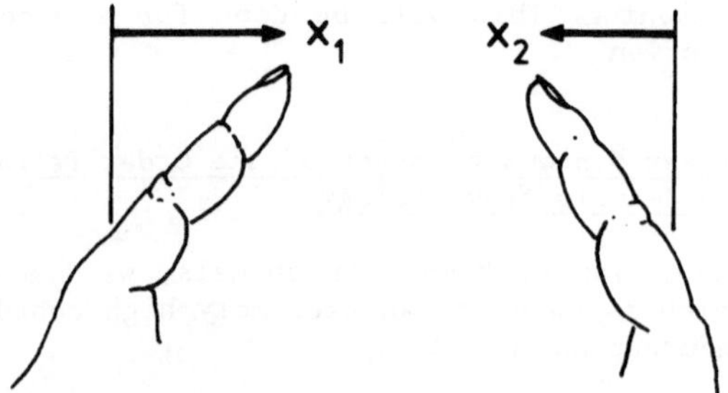

Fig.7: Movement with opposite phases

behavior. To this end we write [9] the elongations of the fingertips in
the form

$$x_1 = r_1 \cos(\omega t + \Phi_1), \qquad\qquad (4.1)$$

$$x_2 = r_2 \cos(\omega t + \Phi_2), \qquad\qquad (4.2)$$

where ω is the basic frequency of the hand movement, while the
amplitudes r_1, r_2 and the phases Φ_1, Φ_2 are time dependent
quantities whose time dependence is assumed to be much slower than that
defined by the frequency ω.

We define the relative phase by

$$\Phi = \Phi_2 - \Phi_1. \qquad\qquad (4.3)$$

In order to describe the change of phase we adopt our basic ideas
introduced above. We may expect that the order parameter equation is of
the form

$$\dot{\Phi} = - \frac{\partial V}{\partial \Phi}, \qquad\qquad (4.4)$$

where V is a potential function in some analogy to that introduced above.
In search for a model we make a few rather obvious assumptions about V.
Since Φ occurs under cosine or sine functions, the properties of a
physical system must not change when Φ is replaced by $\Phi + 2\pi$.
Consequently, we shall postulate that the potential V is periodic:

$$V(\Phi + 2\pi) = V(\Phi). \qquad\qquad (4.5)$$

We furthermore introduce the assumption that both hands play a symmetric
role. In such a case the behavior of the system must not depend on the way
we label the right hand and the left hand. This means, that V must remain
unchanged when we exchange the indices 1 and 2 of the two fingers. This in
turn means that the potential V is symmetric:

$$V(\Phi) = V(-\Phi). \qquad\qquad (4.6)$$

We assume that V obeys the conditions (4.5) and (4.6) in the simplest
possible form which explains the above mentioned experimental results. To
this end we write V as a superposition of two cosine functions:

$$V(\Phi) = - a \cos(\Phi) - b \cos(2\Phi). \qquad\qquad (4.7)$$

As we have seen above the behavior of the system obeying equation (4.4)
can be easily described by identifying Φ with the coordinate of a
particle which moves in an overdamped fashion in the potential V. This
potential V is represented in Fig.8 for various values of the ratio b/a
where we assume that b/a decreases with increasing frequency. Quite
evidently at a critical value ω_c the ball makes a transition from the
state $\Phi = \pi$ to $\Phi = 0$, or equivalently we may say that the
antisymmetric hand movement makes a transition to the symmetric hand
movement.

On the other hand, when we decrease ω starting from high values the
system remains all the time in the $\Phi = 0$ state, even if ω drops below
ω_c. This "hysteresis" phenomenon which easily follows from our simple

model, is also found in the real experiments.

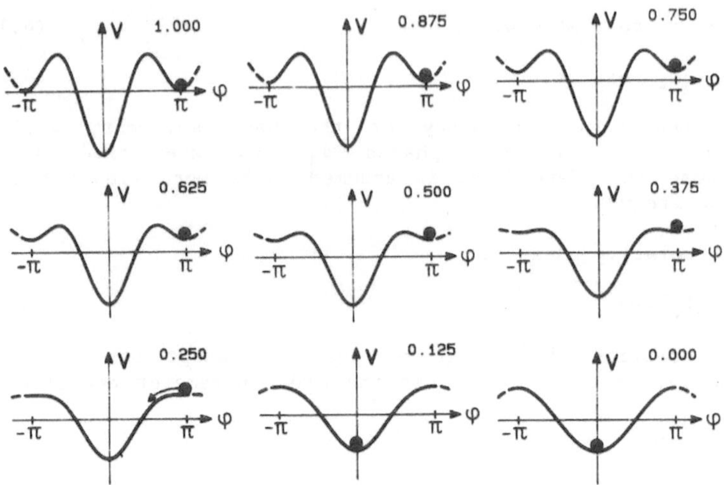

Fig.8: The potential V for different values of b/a

This model has been refined in our original paper [9] to treat the oscillatry motion of the hands explicitly. However, in the context of our lecture we wish still to dwell on the analogy with the phase transition in particular with respect to the phenomenon of critical fluctuations which we have mentioned above.

Looking at our model and having in mind the typical critical fluctuations of synergetic systems close to their transition points, we suggested to Kelso to look for such fluctuations. Fig.9 shows his experimental results.

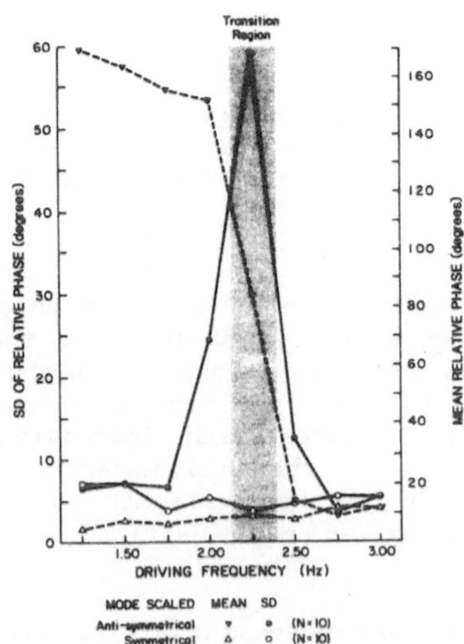

Fig.9: Mean value and standard deviation of Φ as
a function of frequency

In this figure both the average phase and the phase fluctuations, i.e. more precisely the root mean square of the phase fluctuations, are plotted. In the case of the transition from the antisymmetric to the symmetric hand motion has been studied.

Therefore big critical fluctuations occur indeed. We have modelled this transition under the impact of fluctuations [10] by means of adding a fluctuating force to the equation (4.4). I.e., we have treated the equation

$$\dot{\Phi} = -\frac{\partial V}{\partial \Phi} + F(t). \tag{4.8}$$

Using the Fokker-Planck equation, we have studied both the behavior of the root mean square as well as correlation functions and results are found in excellent agreement with the experiment.

What is most interesting and important, is the consequence of this treatment. Before we discuss it we briefly mention that by extension of our model which takes into account the oscillatry motion of the hands, we may reproduce the experimental curve as shown in Fig.10.

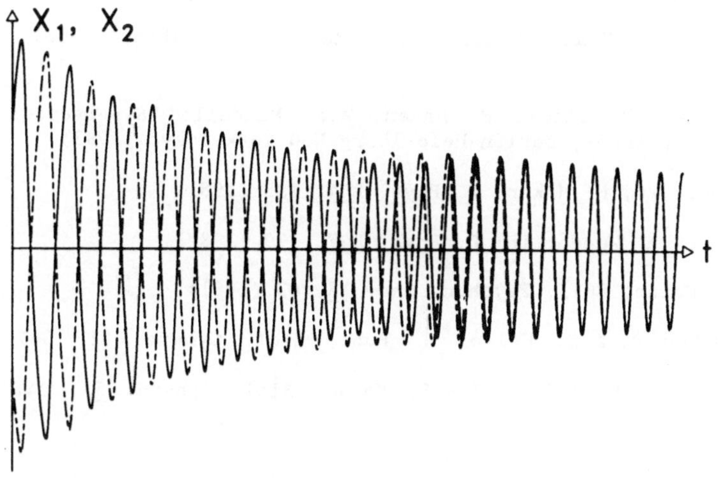

Fig.10: The transition when the frequency is slowly
        raised with time

But let us now discuss the important consequences. Namely when we first assume that the transition between one kind of hand movement to the other kind is caused by the change of the motorprogram of the neurons, it will be very hard to understand why any fluctuations should occur at all. Indeed a motorprogram is a fixed program and no fluctuations should be expected. The way the transition occurs in the hand movement rather indicates that we are dealing here with a typical act of self-organization. This system organizes itself, that means, the individual neuron and muscles act jointly as if the whole system acts as a total autonomous system.

Quite clearly this introduces an entirely new paradigm in biology and it can be hoped that simmilar mechanisms and models apply to more complicated

motions, where the next step will be to study the change of gaits of
horses. Other highly coordinated motions may most probably be treated very
much the same way for instance rhythmic motions like breathing and
heartbeat and their coordination.

References:

[1]   H. Haken: "Synergetics. An Introduction". Springer, Berlin-
Heidelberg, New York (1983).

[2]   H. Haken: "Advanced Synergetics". Springer, Berlin-Heidelberg-
New York-Tokyo (1983).

[3]   H. Haken: "Laser Theory". In Encyclopedia of Physics, Vol.XXV/2c.
Springer, Berlin-Heidelberg-New York (1970).

[4]   e.g. A.M. Zhabotinskii, A.N. Zaikin: J. Theor. Biol. 40, 45
(1973).

[5]   E. Basar, H. Flohr, H. Haken, A.J. Mandell: "Synergetics of the
Brain". Springer, Berlin-Heidelberg-New York-Tokyo (1983).

[6]   A. Wunderlin, H. Haken: Z. Phys. B44, 135 (1981).

[7]   H. Haken, A. Wunderlin: Z. Phys. B47, 179 (1982).

[8]   J.A.S. Kelso: Bull. Psychon. Soc. 18, 63 (1981).

[9]   H. Haken, J.A.S. Kelso. H. Bunz: Biol. Cybern. 51, 347 (1985).

[10] G. Schöner, H. Haken, J.A.S. Kelso: Biol. Cybern. 53, 247 (1986).

Das Osnabrücker Biosphären Modell als Simulationsmodell zur Beschreibung der globalen Änderung des Kohlenstoffkreislaufs

Helmut Lieth, Osnabrück

Zusammenfassung. Das Osnabrücker Biosphären Modell ist ein teilweise dynamisches Simulationsmodell der Biosphäre. Im wesentlichen soll es den globalen Kohlenstoffumsatz zwischen Atmosphäre und Biosphäre in Abhängigkeit von Umweltparametern wie Klima, Boden und menschliche Einflüsse simulieren. Um den globalen Umsatz bilanzieren zu können, ist es notwendig, den Austausch von CO_2 zwischen Ozean und Atmosphäre zu berücksichtigen. Ebenso muß der anthropogene Ausstoß an CO_2 bekannt sein. Das Modell beschreibt Eigenschaften und Funktionen der Biosphäre teilweise genauer als für die C-Bilanzierung notwendig. Dies ist getan worden, weil das Modell in Zukunft für andere Fragestellungen benutzt werden soll.

In der vorliegenden Arbeit wird die Version 1 des Osnabrücker Biosphären Modells besprochen. Da das Modell bereits mehrfach veröffentlicht worden ist (ESSER 1984, LIETH 1984 im Druck, ESSER 1985, LIETH und ESSER 1985, ESSER 1986), soll in dieser Arbeit nur das Gleichungssystem besprochen werden. Die im Modell verwendeten Tabellenfunktionen und die mit dem Modell erzielbaren Ergebnisse können den genannten Arbeiten entnommen werden. Z. Z. wird an der Version 2 des Modells gearbeitet, dessen Eigenschaften am Schluß kurz beschrieben werden. Die neuen Versionen des Modells werden in Kürze von HASSELMANN et al. veröffentlicht werden. Bandkopien der Version 1, einschließlich der notwendigen Datenbank, können beim Autor zum Selbstkostenpreis bestellt werden.

Das Osnabrücker Biosphärenmodell (OBM, Version 1) simuliert den Kohlenstofffluß durch die Biosphäre unter Einschluß des anthropogen erzeugten Kohlendioxids durch Verbrennung fossiler Kohlenstoffquellen und durch Waldrodungen. Der Biomassenaufbau und -abbau der großen Vegetationseinheiten wird in Abhängigkeit von Umweltparametern in Differentialgleichungen und Tabellenfunktionen beschrieben. Mit zwölf Gleichungsgruppen und 5 Tabellen (fossiler Brennstoffverbrauch, Bodenfruchtbarkeit, Vegetationseigenschaften, globale Muster mittlerer Jahresniederschläge und mittlerer Jahrestemperaturen) ist das Biosphärenmodell so aufgebaut, daß es den Kohlenstoff-Fluß von 1860 bis 1980 simulieren kann und dabei nachprüfbare Größen wie den CO_2-Gehalt der Luft sowie die Vorräte an lebender und toter Biomasse hinreichend genau für verschiedene Jahre berechnet.

Dynamisiert wird das Modell durch Änderungen des CO_2-Gehaltes der Luft,

---

1 Die Arbeiten am Osnabrücker Biosphären Modell wurden erst von der DFG, später vom UBA und der EG gefördert. Z.Z. erfolgt Förderung über das *Klimaprogramm des BMFT*. Die Version 1 wurde vorwiegend von der EG unter Nr. CLI-040-D gefördert.

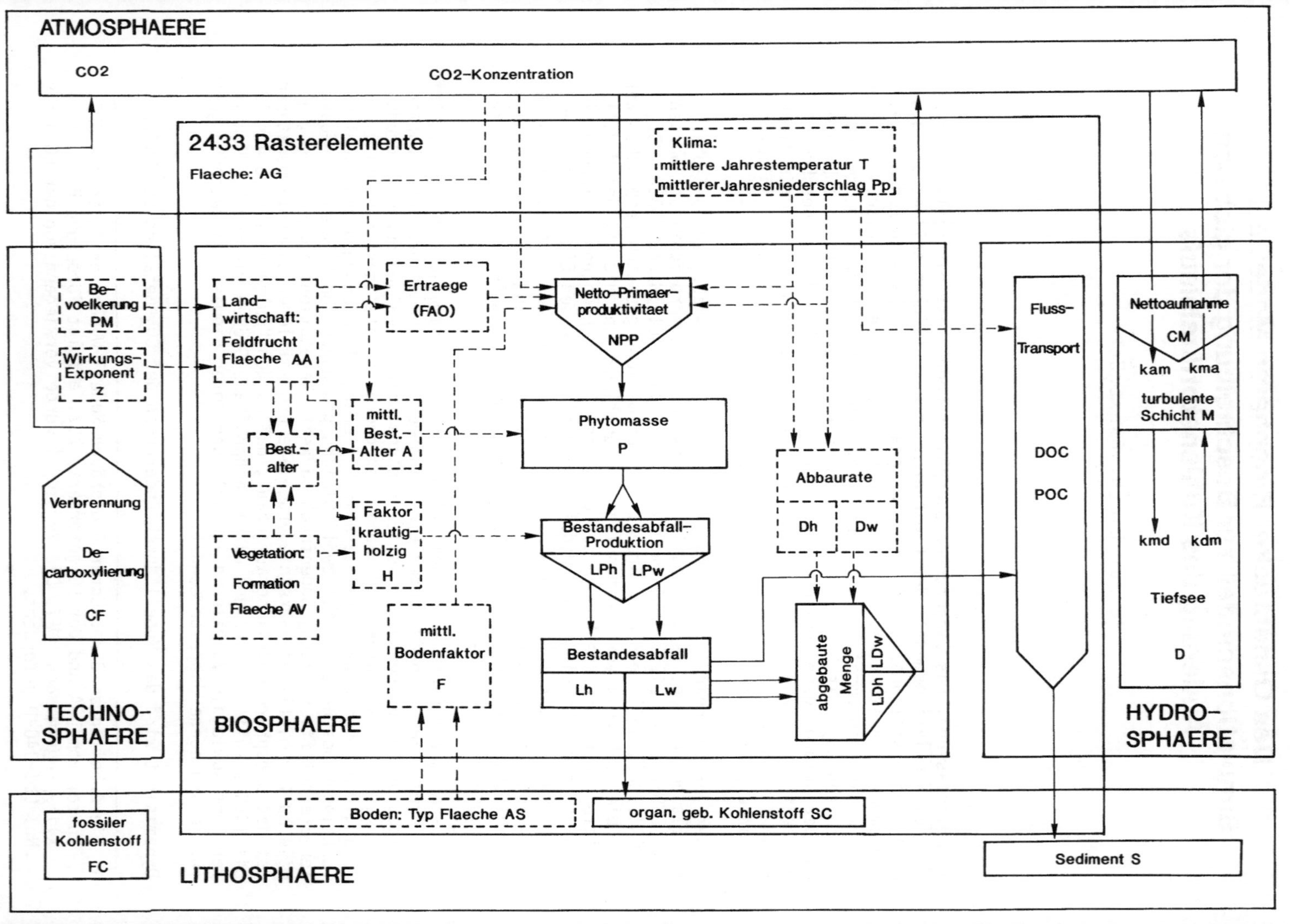

ATMOSPHAERE
CO2
CO2-Konzentration
2433 Rasterelemente
Flaeche: AG
Klima:
mittlere Jahrestemperatur T
mittlerer Jahresniederschlag Pp
Be-voelkerung PM
Wirkungs-Exponent z
Land-wirtschaft: Feldfrucht Flaeche AA
Ertraege (FAO)
Netto-Primaer-produktivitaet NPP
Phytomasse P
mittl. Best.-Alter A
Best.-alter
Faktor krautig-holzig H
Vegetation: Formation Flaeche AV
Bestandesabfall-Produktion
LPh   LPw
mittl. Bodenfaktor F
Bestandesabfall
Lh   Lw
Abbaurate
Dh   Dw
abgebaute Menge
LDh   LDw
Verbrennung
De-carboxylierung CF
Fluss-Transport
DOC
POC
Nettoaufnahme CM
kam   kma
turbulente Schicht M
kmd   kdm
Tiefsee D
TECHNO-SPHAERE
BIOSPHAERE
HYDRO-SPHAERE
fossiler Kohlenstoff FC
Boden: Typ Flaeche AS
organ. geb. Kohlenstoff SC
Sediment S
LITHOSPHAERE
30

durch die wachsende Weltbevölkerung und durch die tabellarisch vorgegebene
anthropogene CO_2-Entwicklung. Mehrere Szenarien werden verglichen und dabei
festgestellt, daß unter der Annahme eines konstanten Vegetationsalters und
einer dynamisch über die wachsende Weltbevölkerung an das Modell gekoppelte
Flußfracht (zusätzliche C-Senke, die nicht über die Atmosphäre in den Ozean
gelangt) das globale Kohlenstoff-Flußsystem zwischen 1860 und 1980 am genau-
esten beschrieben wird.

Einleitung

Die Modellierung des globalen CO_2-Flusses ist von Bedeutung sowohl für das
Klima als auch für eine mögliche Reaktion ganzer Pflanzenbestände auf CO_2-
Konzentrationsänderungen in der Atmosphäre. Siehe hierzu die Arbeiten von
LIETH, FANTECHI, SCHNITZLER eds. (1984) und McCRACKEN und LUTHER (1985).
Ein Modell dieser Art muß daher in der Lage sein, die Wechselbeziehungen
zwischen Atmosphäre, Biosphäre und Ozean in ihrer zeitlichen Dynamik zu be-
schreiben und nach Möglichkeit zu erklären. Diese Aufgabe wurde im Osnabrük-
ker Biosphären Modell angegangen.

Die Abbildung 1 zeigt das statische Flußschema des Osnabrücker Biosphären
Modells (OBM). In ihr sind die für das Modell relevanten Flüsse, Pools und
Parameter genannt, die man quantitativ ermitteln muß. Das sind die Mengen
an CO_2 in der Atmosphäre, in der lebenden Biomasse und im Humus einschließ-
lich des Bestandesabfalls, die Transferraten von Atmosphäre zu Pflanzen (die
Nettoprimärproduktivität, NPP), von lebenden Pflanzen zu Humus, Humus zu At-
mosphäre (der Abbau, teilweise als Bodenatmung bezeichnet), von Biosphäre zu
Atmosphäre durch den Eingriff des Menschen bei land- und forstwirtschaftli-

---

Legende zu Abb. 1: Flußschema für das Osnabrücker Biosphären Modell

Bedeutung der Symbole:

 □   Pools   ⟶ Massenbeziehung

        - - -► Regelbeziehung

 ▽   Massenströme

 ⌐¬   Regelvariablen
 └┘

Gleichungen:

Erhaltung der Masse

$$1. \quad \sum_{k=1}^{n} \frac{d\,\square_k}{dt} = 0 \qquad\qquad 2. \quad \frac{d\,\square_k}{dt} = \sum_{i=1}^{m(k)} \triangledown_{k,i} \quad ; \ k=1,\ldots,n$$

Modellfunktionen

$$3. \quad \triangledown_{k,i} = f_{k,i}(\ulcorner\,\urcorner_1 ,\ldots, \ulcorner\,\urcorner_j , \square_1 ,\ldots, \square_n) \quad ; \ k=1,\ldots,n$$
$$i=1,\ldots,m(k)$$

cher Nutzung, Geosphäre zu Atmosphäre durch den CO_2-Ausstoß beim Verbrennen
fossiler und rezenter organischer Brennstoffe, Atmosphäre zu Ozean durch die
temperaturabhängigen Gleichgewichte und schließlich Biosphäre zu Ozean, be-
dingt durch die Flußfrachten sowie Ozean zu Geosphäre, als Folge der Sedi-
mentation.

Die genannten Zwischensenken, zeitlich begrenzte Anhäufung von lebender oder
toter Biomasse, die in den Simulationsmodellen als Pools bezeichnet werden,
und die Umsatzgeschwindigkeiten, die von den Modellierern Flußraten genannt
werden, muß der Geoökologe, soweit sie die Biosphäre betreffen, als regio-
nales Muster so genau wie möglich erfassen. Die dazu benötigten Datensätze
werden deshalb zunächst vorgestellt.

Die notwendige Datenbasis

Ein Modell dieser Art benötigt eine große Zahl von Daten über die Vegetation und über die Einflüsse des Menschen. Die Qualität dieser Daten entschei-
det mit über die spätere Brauchbarkeit des Modells. Die Daten werden am
praktischsten in Tabellenform als Zeit- oder Zustandgrößen-Liste aufgestellt
und können dann in dieser Form in ein EDV-Programm eingegeben werden. Die
auf den folgenden Seiten geschriebenen Tabellen 1 - 4 enthalten die Werte
oder Angaben, die man in Tabellenform für die Version 1 des OBM benötigt.

Neben den genannten Tabellen müssen noch einige Reservoiregrößen bekannt
sein: Das Volumen der Atmosphäre und des Ozeans, die Anzahl von Menschen
auf der Erde 1969, die Kohlenstofffrachten der Flüsse um 1970 sowie für bestimmte Aufgaben der Vorrat an Kohlenstoff in verbrennbaren Lagerstätten.

Diese Daten sind so organisiert, daß sie bei Bedarf vom Hauptprogramm aufge-
rufen und benutzt werden können.

Die Simulation des Kohlenstoff-Flusses durch die Biosphäre

Der Grundgedanke für die Simulation ist der Versuch, Aufbau und Abbau der
biosphärischen Trockensubstanz in Jahresschritten zu bilanzieren. Die notwendigen Berechnungen sind in den folgenden Gleichungsgruppen dargestellt:

Aus den unter (1) genannten Gleichungen wird die Nettoprimärproduktivität
(NPP),das ist die Menge an Trockensubstanz, die je m^2 und Jahr gebildet wird,
wie angegeben, aus den Tabellenwerten T, N, F für Temperatur und Niederschlä-

ge sowie für die Bodenfruchtbarkeit berechnet (siehe Tabellen 2 und 4).

Tabelle 1: Jährlicher anthropogener Kohlenstoff-Ausstoß von 1860 - 1981 nach KEELING (1973, 1982), MARLAND und ROTTY (1983) aus ESSER (1985).

Jahr	10^{12} g C/a	Jahr	10^{12} g C/a	Jahr	10^{12} g C/a
1860	93.3	1900	524.9	1940	1 300.4
1861	98.7	1901	540.3	1941	1 337.1
1862	98.4	1902	552.9	1942	1 334.4
1863	106.0	1903	606.4	1943	1 364.0
1864	115.1	1904	613.4	1944	1 352.2
1865	121.9	1905	646.6	1945	1 203.6
1866	128.7	1906	696.1	1946	1 270.5
1867	137.9	1907	771.2	1947	1 421.5
1868	136.7	1908	736.6	1948	1 517.5
1869	141.8	1909	769.0	1949	1 469.6
1870	145.0	1910	804.8	1950	1 613.0
1871	161.9	1911	821.8	1951	1 743.5
1872	175.9	1912	866.2	1952	1 766.5
1873	188.4	1913	929.0	1953	1 808.1
1874	183.3	1914	838.4	1954	1 880.0
1875	189.2	1915	830.8	1955	1 895.0
1876	191.6	1916	894.8	1956	2 079.0
1877	196.0	1917	945.3	1957	2 215.0
1878	196.5	1918	932.0	1958	2 292.0
1879	207.6	1919	828.9	1959	2 341.0
1880	227.1	1920	958.9	1960	2 457.0
1881	244.2	1921	828.0	1961	2 563.0
1882	262.5	1922	890.6	1962	2 636.0
1883	280.0	1923	1 005.3	1963	2 786.0
1884	282.1	1924	998.5	1964	2 932.0
2885	276.4	1925	1 006.4	1965	3 121.0
1886	278.7	1926	1 006.0	1966	3 247.0
1887	297.7	1927	1 097.5	1967	3 416.0
1888	321.9	1928	1 090.9	1968	3 465.0
1889	328.5	1929	1 171.9	1969	3 688.0
1890	349.7	1930	1 077.5	1970	3 896.0
1891	365.4	1931	968.2	1971	4 245.0
1892	368.7	1932	873.8	1972	4 400.0
1893	361.1	1933	918.8	1973	4 560.0
1894	377.1	1934	996.5	1974	4 797.0
1895	398.6	1935	1 031.7	1975	4 880.0
1896	411.6	1936	1 146.4	1976	4 856.0
1897	431.5	1937	1 226.2	1977	5 064.0
1898	454.6	1938	1 161.4	1978	5 263.0
1899	497.3	1939	1 232.9	1979	5 246.0
				1980	5 170.0
				1981	5 060.0

Tabelle 2: Die Faktoren für $F_{(Bodenfruchtbarkeit)}$ für die häufigsten Böden der Erde nach ESSER (1984) verändert. Die Bodensystematik folgt der Klassifikation der FAO-UNESCO Weltbodenkarte (1974). Die Faktoren werden in der Gleichungsgruppe (1) verwendet.

FAO Bodeneinheiten	Faktor	Anzahl der Datenpunkte
Acrisole		
Ag	0.87	1
Ah	0.22	2
Ao	0.70	7
other A	0.60	–
Cambisole		
Bd	0.94	10
Be	1.69	7
Bh	1.58	3
Bx	0.76	1
Chernoseme		
Cl	0.99	1
Podsoluvisole		
Dd	0.83	1
Ferralsole		
Fx	0.55	1
Gleysole		
Gh	0.47	2
Gx	0.57	2
other G	0.50	–
Lithosole		
I	0.52	7
Iy	1.14	1
Fluvisole		
J	0.49	1
Je	0.61	2
other J	0.55	–
Kastanozeme		
Kh	1.96	2
Kl	1.61	3
other K	1.80	–

FAO Bodeneinheiten	Faktor	Anzahl der Datenpunkte
Luvisole		
La	0.34	1
Lc	1.04	2
Lf	1.65	1
Lg	2.78	1
Lo	0.85	4
Histosole		
Od	1.39	2
Podzole		
Ph	0.56	1
Po	0.61	7
other P	0.55	–
Regosole		
Rc	1.61	2
Re	1.14	1
Rx	0.91	1
other R	1.20	–
Solonetz		
So	0.59	1
Andosole		
Tv	1.65	1
Xerosole		
Xh	0.42	1
Yermosole		
Y	0.30	2
Yh	0.66	1
Yt	0.09	1
Yl	0.23	2
Solonchak		
Zo	0.44	2
Zt	0.03	1
other Z	0.20	–

Tabelle 3: Angenommenes Bestandesalter für das Gleichgewicht
von Trockensubstanzaufbau und -abbau sowie der Anteil an krau-
tiger Biomasse von 31 Formationen der potentiellen natürlichen
Vegetation, die im OBM unterschieden werden. Die Werte wurden
durch Auswertung des Informationssystems DATAVW (ESSER et al.
1982) gewonnen. Fehlten Werte, so wurden die Formationen von
uns eingestuft ("Ranking"). Sie tragen in diesem Fall einen *.
Der Anteil der Vegetationstypen je Rasterfläche (2.5°x2.5°) ist
digitalisiert und als Tabelle ebenfalls vorhanden (siehe Angaben in Tab. 4).

Formation	mittlere Bestandes- alter	Faktor für krau- tigen Anteil an der Gesamt-NPP
TROPISCHE TIEFLANDS-REGENWÄLDER	200	0.37
TROPISCHE TIEFLANDS-TROCKENWÄLDER	80 *	0.4 *
TROPISCHE BERGWÄLDER	80 *	0.37 *
TROPISCHE SAVANNEN	5	0.98
TROPISCHE PARAMO-GEHÖLZE	10 *	0.95 *
TROPISCHES PARAMO-GRASLAND	1 *	1.0 *
PUNA-STEPPEN	2 *	1.0 *
SUBTROPISCHE IMMERGRÜNE WÄLDER	200	0.37
SUBTROPISCHE LAUBABWERFENDE WÄLDER	150	0.44
SUBTROPISCHE SAVANNEN	5	0.90
SUBTROPISCHE HALOPHYTENFORMATIONEN	5	0.90
SUBTROPISCHE STEPPEN UND GRASLÄNDER	1	1.0
TEMPERIERTE STEPPEN UND WIESEN	1	1.0
SUBTROPISCHE HALBWÜSTEN	15	0.85
XEROMORPHE FORMATIONEN	20	0.4
WÜSTEN (TROPISCH, SUBTROPISCH, POLAR)	5 *	0.85 *
MEDITERRANE HARTLAUB-WÄLDER	100 *	0.4
MEDITERRANE GEHÖLZE UND STRAUCHFORMATIONEN	15	0.47
TEMPERIERTE IMMERGRÜNE WÄLDER	130	0.29
TEMPERIERTE LAUBABWERFENDE WÄLDER	150	0.38
TEMPERIERTE GEHÖLZE	25	0.53
TEMPERIERTE STRAUCHFORMATIONEN	10 *	0.85
KÜHLGEMÄSSIGTE MOORE	5	0.48
BOREALE IMMERGRÜNE NADELWÄLDER	100	0.34
BOREALE LAUBABWERFENDE WÄLDER	100	0.38 *
BOREALE GEHÖLZE	15	0.6 *
BOREALE STRAUCHFORMATION	10 *	0.85 *
STRAUCHTUNDREN	10	0.7
KRAUTTUNDREN	2	1.0
AZONALE FORMATIONEN	5 *	0.60 *
MANGROVE	50	0.29

Tabelle 4: Angaben, die in Tabellenform in einen Modellauf eingelesen werden und die sich wegen ihres Umfangs hier nicht wiedergeben lassen. Die Tabellen sind im Datensatz enthalten, der zusammen mit dem OBM vom Autor bestellt werden kann. Ein Ausdruck der Tabellen kann von uns unter der Bezeichnung DATA VW bezogen werden.

- Mittelwerte für die Lufttemperatur je Rastereinheit.

- Mittelwerte für jährliche Niederschlagssummen je Rastereinheit.

- Prozentsätze aller in Tabelle 3 genannten Vegetationstypen für alle

  Rastereinheiten.

- Der Anteil der landwirtschaftlich genutzten Fläche im Jahre 1960 je

  Rastereinheit.

- Die Flächengröße jeder Rastereinheit

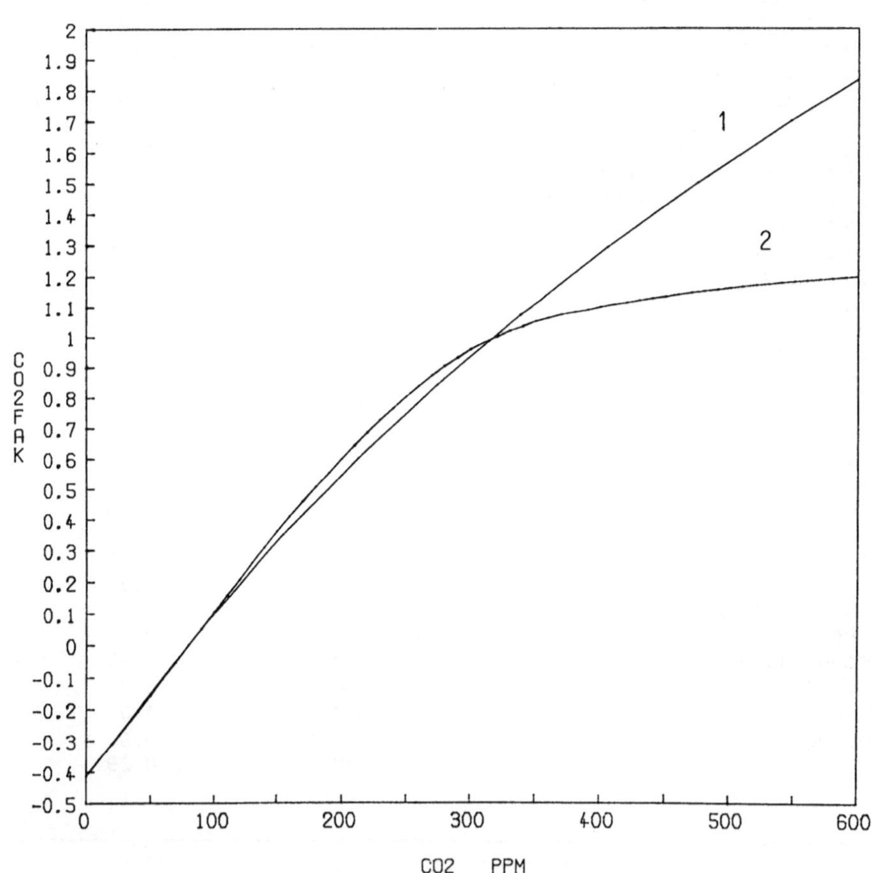

Abbildung 2: Der atmosphärische Düngefaktor F (CO_2) nach Gleichung (2) für fruchtbare Böden (1) und mit einer Spline-Matrix für weniger fruchtbare Böden (2) berechnet. Abszisse = CO_2-Gehalt der Atmosphäre, Ordinate = CO_2-Faktor.

Die Nettoprimärproduktivität (NPP) der Pflanzendecke wird mit folgendem Gleichungssystem berechnet:

$$NPP_{(T)} = 3000/(1+e^{1.315-0.119 \cdot T}) \text{ für Jahresmitteltemperatur}$$

$$NPP_{(N)} = 3000 \cdot (1-e^{-0.0006664 \cdot N}) \text{ für Jahresniederschlagssumme}$$

(1)

$$NPP = \min (NPP (T), NPP (N)) \cdot F_{(Bodenfruchtbarkeit)} \cdot F_{(CO_2)}$$

$F_{(Bodenfruchtbarkeit)}$ wird aus der Tabelle 2 entnommen.

Die NPP wird außerdem vom CO_2-Gehalt der Atmosphäre und anderen anthropoge-
nen Stoffen, die über die Atmosphäre in die Ökosysteme gelangen, beeinflußt.
Wir bezeichnen diesen Wirkungsfaktor als $F_{(CO_2)}$. Er wird in der Version 1
des OBM mit folgender Gleichung berechnet:

(2) $F_{(CO_2)} = 3.6365 \cdot (1-e^{\hat{}}(-0.00135 \cdot (CO_2/ppm-80)))$

Diese Gleichung ist inzwischen für die Version 2 modifiziert worden, da sich
herausgestellt hat, daß sie nur für Regionen mit fruchtbaren Böden gilt. Für
weniger fruchtbare Böden wurde eine einfache Spline-Matrix aufgestellt; die
Werte für $F_{(CO_2)}$ nach Gleichung (2) und der Spline-Matrix sind in Abbildung
2 dargestellt.

Der CO_2-Gehalt der Atmosphäre wird für 1860 mit 260 bis 280 ppmv angenommen
und für 1981 mit 340 ppmv. Die dazwischen liegenden Jahreswerte werden vom
Modell errechnet.

Abbildung 3 zeigt das Produktivitätsmuster der Erde in einem 2.5° Raster berechnet mit der Gleichungsgruppe 1 unter der Annahme $F_{(CO_2)} = 1$.

Aus den NPP-Werten wird die Phytomasse errrechnet unter der Annahme, daß die
Vegetationstypen bis zu den Endwerten des in Tabelle 3 angegebenen Bestandes-
alters entsprechend der Gleichung (3) zunehmen.

Phytomasse im "steady state" Bestandesalter (A) und mittlere Produkti-
vität (NPP).

(3) $P/kg = 0.5918 \cdot 10^{\hat{}}-3 \cdot A^{\hat{}}(0.79216 \cdot NPP)$

Krautige Vegetation erhält für das 2. Jahr bereits den Höchstwert, der mit
der jährlichen NPP gleichgesetzt wird. Alle landwirtschaftlich genutzten

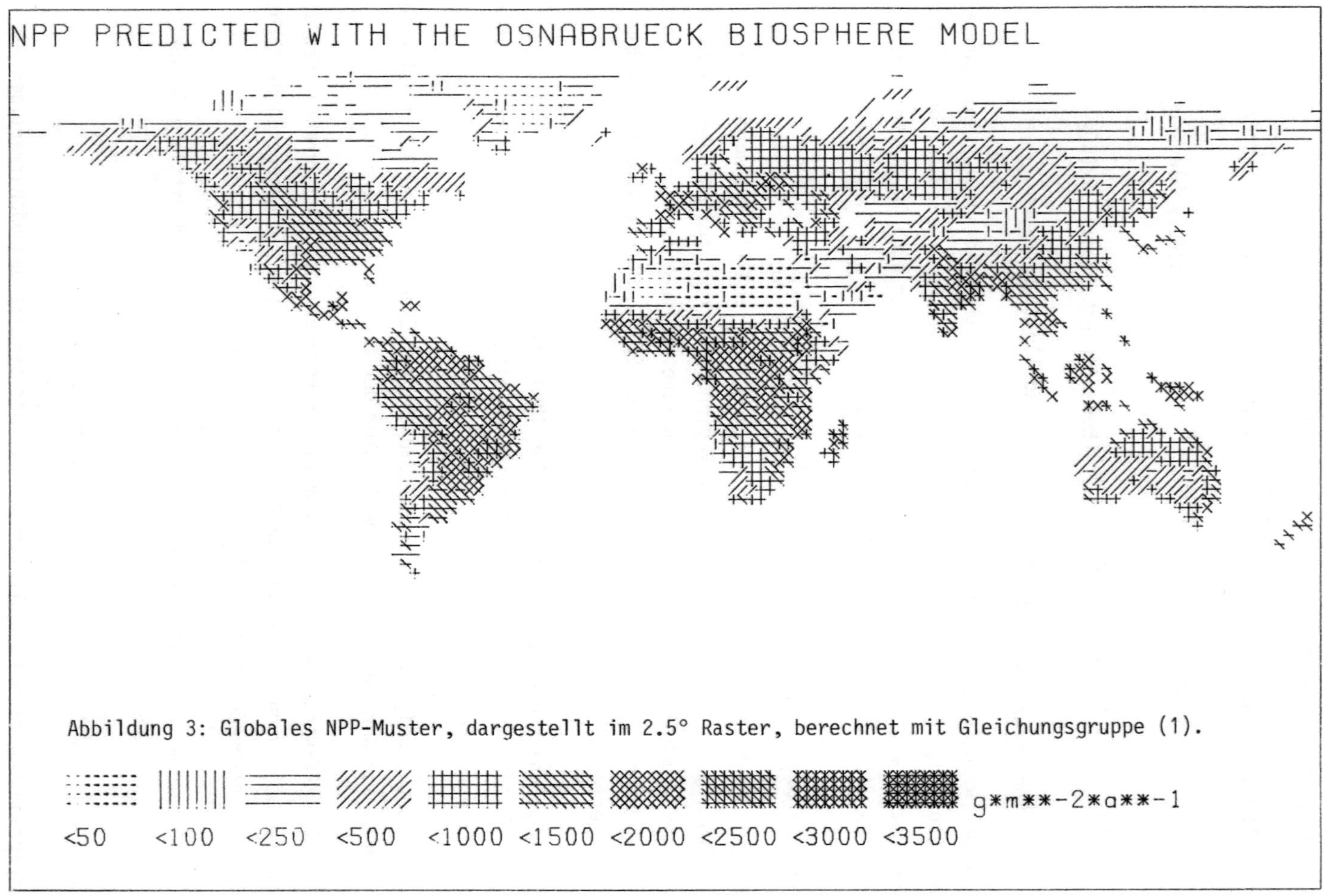

Abbildung 3: Globales NPP-Muster, dargestellt im 2.5° Raster, berechnet mit Gleichungsgruppe (1).

Flächen gehören dazu, und deshalb ist es wichtig zu wissen, wie sich die
Kulturflächen im Laufe der Zeit verändert haben. Die Fläche ist für 1970 aus
der FAO-Statistik bekannt und ist in unsere Flächenmatrix länderweise über-
tragen worden. Für jede Fläche wird bis zur Erschöpfung des Areals natürli-
che Vegetation entsprechend dem Bevölkerungswachstum als gerodete Fläche angenommen. Die Berechnung geht nach der Gleichungsgruppe (4).

Landwirtschaftliche Fläche (AA) für das Jahr (I) beträgt

$$AA_{(I)} = AA\ (1970) \cdot \left(\frac{PM_{(I)}}{PM\ (1970)} \right)^{K}$$

(4)

$$PM_{(I)} = \text{Globale Bevölkerung} = 3.9371 \cdot 10^7 \cdot e^{(0.0205 \cdot (I-1760))} + 7.297 \cdot 10^8$$

I = Jahr zwischen 1760 und 1981

K = Szenarioexponent, der willkürlich gesetzt wird zwischen o.8 und 1.
Der Faktor soll die Effizienz angeben, mit der die Weltbevölkerung
die landwirtschaflich genutzte Fläche verwendet.

In der Version 2, die sich in Arbeit befindet, wird die Gleichungsgruppe (4)
durch eine Tabellenfunktion ersetzt.

Aus der jährlich produzierten und stehenden Phytomasse wird Bestandesabfall,
der je nach der in Tabelle 3 ausgewiesenen Kategorie krautig oder holzig
früher oder später anfällt. Im "steady state" Bestandesalter spielt das jedoch keine Rolle mehr, dann wird die ganze NPP als Bestandesabfall angenommen und dieser entsprechend seiner Beschaffenheit, krautig oder holzig, un-
terschiedlich schnell abgebaut.

Der jährlich anfallende Bestandesabfall (LP) berechnet sich nach Gleichungsgruppe (5)

$$LP\ /g \cdot m^{-2} \cdot a^{-1} = NPP - P - E$$

(5.1)

P = jährliche Phytomassenänderung

E = geerntete Biomasse

Für (LP) wird entsprechend der Tabelle 3 und der hier nicht wiedergegebenen
Digitalisierung der Vegetationseinheiten für jede Rasterfläche die Gesamt-
menge an krautigem und holzigem Bestandesabfall berechnet.

$$L(i)/kg = (\ \sum_{1}^{2433}\ (Lh(i,m)\cdot AG(m))\ +\ \sum_{1}^{2433}\ (Lw(i,m)\cdot AG(m))\ )\ \cdot 10^3$$

$$
\begin{aligned}
i &= \text{Jahr } i\\
(5.2)\quad m &= \text{Rasterelement } m\\
AG(m) &= \text{Rasterflächengröße in } Km^2
\end{aligned}
$$

Die angefallene Menge an Holz und Kraut wird entsprechend den klimatischen
Bedingungen durch Mikroben zersetzt und dabei der biologisch gebundene Koh-
lenstoff wieder zu CO_2 remineralisiert. Das geschieht im Modell nach der
Gleichungsgruppe (6)

Litter-Abbaurate in % des holzigen und krautigen Litterpools pro Jahr

Krautig

$$Dh/\%\cdot a^{-1} = \min(F1\ (T),\ G1\ (N))$$

$$(6.1)\quad F1(T) = 7.67\cdot e^{(0.0926\cdot(T/°C+6.41))} + 17.6$$

$$G1(N) = (\frac{50}{0.215+e^{(4.2-0.0053\cdot N/mm)}} + 670)\cdot(\frac{0.094}{0.7+e^{(0.0023\cdot N/mm-5.05)}}$$

$$+0.076)\cdot 0.64\cdot(1-e^{(-0.001\cdot N/mm)})$$

Holzig

$$Dw/\%\cdot a^{-1} = \min\ (F2\ (T),\ G2\ (N))$$

$$(6.2)\quad F2(T) = 2.67\cdot e^{(0.0522\cdot(T/°C+31.63))} - 2.51$$

$$G2(N) = (\frac{27.8}{0.021+e^{(8.53-0.0095\cdot N/mm)}} + 712)\cdot(\frac{0.126}{1.51+e^{(0.003\cdot N/mm-4.65)}}$$

$$+0.05)\cdot 0.5\cdot(1-e^{(-0.001\cdot N/mm)})$$

$$
\begin{aligned}
T/°C &= \text{Mittlere Jahrestemperatur}\\
N/mm &= \text{Mittlere jährliche Niederschlagssumme}\\
Dh &= \text{Abbaurate krautig}\\
Dw &= \text{Abbaurate holzig}
\end{aligned}
$$

Aus den Gleichungen (5.2) und (6) wird der jährliche Biomassenzerfall (LD)
je Rasterfläche berechnet. Das Ergebnis geht in die globale Bilanzrechnung
für den jährlichen Kohlenstoffumsatz der Biosphäre ein. Dabei wird angenom-

men, daß 45 % der biologischen Trockensubstanz aus Kohlenstoff besteht. Gleichung (7) erläutert den Rechenvorgang:

$$C = \sum_{1}^{2433} C(i,m)/g \cdot m^{-2} \cdot a^{-1} = 0.45 \cdot 10^{-6} \cdot \sum_{1}^{2433} (NPP(i,m) - LD(i,m)) \cdot AG(m)$$

(7)
$$i \quad = \text{Jahr } i$$
$$m \quad = \text{Rasterelement } m$$
$$AG(m) = \text{Rasterflächengröße in } Km^2$$

Die Berechnung der globalen Kohlenstoffbilanz

Die Teilrechnungen aus der Biosphäre müssen in die globale Kohlenstoffbilanz eingebaut werden. Dazu brauchen wir zusätzlich den fossilen Brennstoffverbrauch je Jahr nach Tabelle 1 (MARLAND und ROTTY, 1983) und den Kohlenstoffaustausch zwischen Atmosphäre und Ozean, der in Version 1 entsprechend der Gleichungsgruppe (8) berechnet wird (SIEGENTHALER und OESCHGER 1978).

$$F_{am} \quad = K_{am} (N_n + na) - K_{ma} (N_m - \xi \cdot nm)$$
$$nm \quad = F_{am} - F_{mt}$$
$$F_{mt} \quad = K_{mt} (N_n + nm) - K_{tm} (N_a - nt)$$
$$N \quad = Gt \ C \ 1860$$

(8)
$$n \quad = \text{Zunahme in Gt C seit 1860}$$
$$a \quad = \text{Atmosphäre}$$
$$m \quad = \text{Mischungsschicht des Ozeans}$$
$$t \quad = \text{Tiefsee}$$

$$K_{am} = \frac{1}{7.53} \quad K_{ma} = \frac{1}{9.73} \quad K_{mt} = \frac{1}{22.7} \quad K_{tm} = \frac{1}{1127} \quad \xi = 10$$

Der Anteil an jährlich verbranntem fossilen Kohlenstoff, der in den Ozean geht, wird zu etwa 30 % angenommen. Zusätzlich dazu wird eine bestimmte Menge an Kohlenstoff durch die Flüsse ins Meer getragen, der im Biosphärenmodell nicht erfaßt wird. Er wird im OBM Version 1 für 1970 mit einer Gigatonne angenommen und für die Laufzeit 1860 - 1981 dynamisch an das Bevölkerungswachstum gekoppelt.

Die CO_2-Menge, die jährlich anfällt oder verschwindet, wird mit Gleichung
(9) berechnet.

$$(9) \quad \Delta CO_2/\overline{m}mp = (\ \sum_{1}^{2433}\ C(i,m)/g - CF(i)/g) + Fam(i)/g) \cdot 4.7357 \cdot 10^{-16}$$

CF(i)/g = Gesamtverbrauch fossilen Kohlenstoffs im Jahr i
i = Jahr i
m = Rasterelement m
Fam/g = C - Abgabe der Atmosphäre ins Meer

Wenn man den Modell-Lauf 1860 mit 260 - 290 ppm CO_2 in der Atmosphäre startet, dann erreicht man, je nach Szenario, 1981 etwa 340 ppm CO_2 (ESSER 1985)
Das Ergebnis verschiedener Läufe ist in Abbildung 4 dargestellt. Tabelle 5
gibt für das Szenario ●—●, das am besten die Messwerte am Mauna Loa Observatorium simuliert, die Biosphärenwerte in Dekadenabständen an.

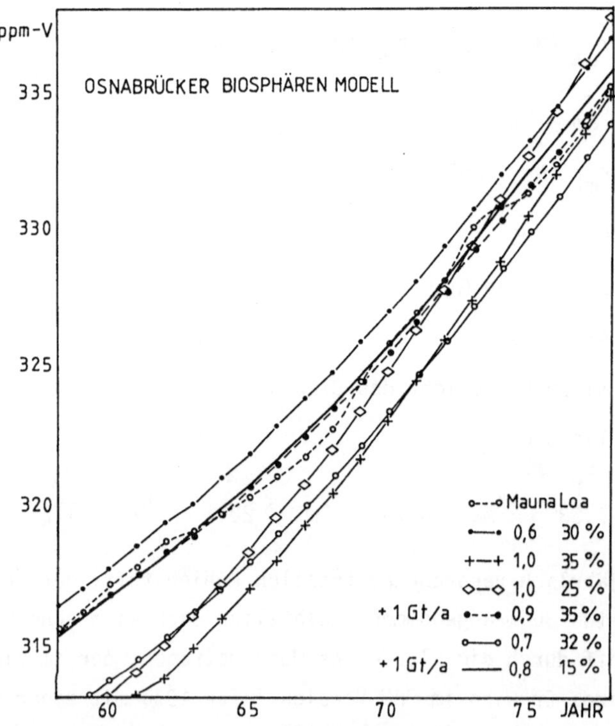

Abbildung 4: CO_2-Konzentration in der Atmosphäre von 1860 - 1981 berechnet
mit dem Osnabrücker Biosphären Modell (aus LIETH und ESSER 1985), verändert.

Tabelle 5: Entwicklung der Pools des Osnabrücker Biosphären Modells bei Annahme einer zusätzlichen, globalen, dynamisch an die Entwicklung der Land-
nutzung gekoppelten Senke von 1 Gt C für das Jahr 1970 (Spalte 3) für den
Zeitraum 1860 bis 1981. Die Senke entspricht den Schätzungen für die Flußfrachten an organisch gebundenem Kohlenstoff (KEMPE 1984). Angaben in Gt C.

Jahr	CO_2 ppm	Flussfrachten pro Jahr	Flussfrachten akkumu-liert	Phyto-masse gesamt	Bestandes-abfall	Rodungen seit 1860	Ozean Zunahme seit 1860
(1)	(2)	(3)	(4)	(5)	(6)	(7)	(8)
1860	295.0	0.37	0.37	630.9	141.3	-0.0	0.0
1870	295.0	0.38	4.1	627.4	142.3	-3.3	0.0
1880	295.3	0.41	8.1	624.3	142.4	-7.2	0.0
1890	295.9	0.43	12.3	621.2	142.5	-12.0	0.3
1900	297.0	0.47	16.9	618.0	142.7	-17.8	0.7
1910	298.7	0.51	21.8	615.2	143.1	-24.8	1.4
1920	300.9	0.57	27.2	612.1	143.9	-33.3	2.5
1930	303.6	0.62	33.2	608.3	144.7	-43.5	4.0
1940	306.7	0.69	39.8	603.0	145.9	-55.9	5.8
1950	310.7	0.78	47.2	597.1	147.3	-70.9	8.3
1960	316.7	0.88	55.5	591.8	148.7	-89.1	11.8
1970	325.7	1.00	64.9	587.9	150.8	-111.2	16.9
1980	338.7	1.14	75.7	587.7	153.9	-137.1	24.2
1981	340.0	1.16	76.8	587.5	154.4	-139.9	25.0

Ausblick auf zukünftige Arbeiten

Das Osnabrücker Biosphären Modell läßt sich ausgezeichnet zur Simulation
des globalen Kohlenstoffkreislaufes verwenden. Hierbei werden in der Version
1 noch einige Annahmen gemacht, die durch Tabellenfunktionen ersetzt werden
müssen, wie z.B. die Entwicklung der landwirtschaftlichen Anbauflächen bzw.
Module, die durch bessere ersetzt werden müssen, wie z.B. das Ozeanmodul.
Das in Version 1 verwendete Ozeanmodul nach SIEGENTHALER und OESCHGER (1978)
soll schrittweise durch andere, z.B. das von MAIER-REIMER (1984) ersetzt
werden.

An den Verbesserungen wird z.Z. gearbeitet. Vorgesehen ist auch die Erweiterung des Modells auf die Simulation monatlicher Biosphärenaktivitäten. Dadurch könnten auch die von der Referenzmeßstelle am Mauna Loa gefundenen jahreszeitlichen CO_2-Schwankungen mit simuliert werden.

<u>Literatur</u>

ESSER, G., I. ASELMANN & H. LIETH (1982): Modelling the carbon reservoir in the system compartment "litter". In: (E.T. Degens, ed.) "Transport of Carbon and Minerals in Major World Rivers, Vol. 1". Mitt. Geol.-Paläont. Inst. Univ. Hamburg, SCOPE/UNEP Sonderbd. 52, 39-58.

ESSER, G. (1984): The significance of biospheric carbon pools and fluxes for atmospheric CO_2: A proposed model structure. Progress in Biometeorology <u>3</u>, 253-294.

ESSER, G. (1985): Der Kohlenstoff-Haushalt der Atmosphäre - Struktur und erste Ergebnisse des Osnabrücker Biosphären Modells. Veröff. Naturf. Ges. zu Emden von 1814. Bd. 15, 160 S. mit 27 Abb. (Habilitationsschrift).

ESSER, G. (1986): Sensitivity of global carbon pools and fluxes to human and potential climatic impacts. Tellus (im Druck).

FAO/UNESCO (1974 ff.): Soil Map of the World. Vol. I-X, Paris.

HASSELMANN, K., H. LIETH, E. MAIER-Reimer, G. ESSER (1986): Interactions between Ocean and terrestrial Biosphere with respect to atmospheric CO_2-variations. Tellus (im Druck).

KEELING, C.D. (1973): Industrial production of carbon dioxide from fossil fuels and limestone. Tellus <u>25</u>, 174-197.

KEELING, C.D. (1982): The global carbon cycle: What we know and could know from atmospheric, biospheric, and oceanic information. Proceedings, CO_2 research conference: Carbon dioxide, science, and consensus. Carbon Dioxide Research Division, U.S. Dept. of Energy. Washington, II. 63 - II. 75.

KEMPE, S. (1984): Sinks of the anthropogenically enhanced carbon cycle in surface fresh waters. J. Geophys. Res. Vol. 89, No. D3, 46, 4657-4676.

LIETH, H., R. FANTECHI, H. SCHNITZLER, eds. (1984): Interaction between Climate and Biosphere. Progress in Biometeorology, Vol. 3, Swets & Zeitlinger, Lisse, Niederlande.

LIETH, H. (1984): Die Rolle der Biosphäre für den globalen CO_2-Haushalt. Tagung vom 29.11. bis 1.12.84 "Atmosphärischer Treibhauseffekt und Klimaentwicklung" DDR, Halle (Leopoldina) (im Druck).

LIETH, H. G. ESSER (1985): The attempt to simulate the global carbon flux from 1860 to 1981 using the Osnabrück Biosphere Model. Mitt. Geol.-Paläontolog. Inst. Univ. Hamburg, SCOPE/UNEP Sonderbd., Heft 58, 137-144.

MAIER-REIMER, E. (1984): Towards a global ocean carbon model. Progress in Biometeorology, Vol. 3, 295-311.

MARLAND, G., R.M. ROTTY (1983): Carbon dioxide emissions from fossil fuels:
    A procedure for estimation and results for 1950-1981. Report DOE/NBB-
    0036 for United States Dept. of Energy, Washington, D.C.

MCCRACKEN, M.C., F.H. LUTHER (eds.) (1985): Detecting the climatic effects
    of increasing carbon dioxide (DOE/ER-0235). U.S. Dept. of Energy,
    Washington, D.C. Available from NTIS, Springfield, Virginia.

SIEGENTHALER, U., H. OESCHGER (1978): Predicting future atmospheric carbon
    dioxide levels. Science 199, 388-395.

A Dynamic Simulation Model of Tree Development under Pollution Stress

Hartmut Bossel

Zusammenfassung. Mittels eines dynamischen Simulationsmodells der wichtig-
sten Prozesse im Baum wurde die Wachstumsdynamik bei Schadstoffbelastung der
Blätter und/oder Feinwurzeln mit Daten für Fichte untersucht. Das Modell
enthält die jahreszeitlichen Veränderungen und berechnet die Blattmasse jeder
Nadel-Altersklasse getrennt als Funktion der Schadstoffbelastung. Bei feh-
lender Umweltbelastung ergeben sich die normalen Wachstumsdaten für Fichte.
Bei Schadstoffbelastung zeigen sich zwei verschiedene Verhaltensmodi: bei un-
terkritischer Belastung überlebt der Baum, stagniert aber im Wachstum; bei
überkritischer Belastung ergibt sich früher oder später ein plötzlicher Zusammenbruch.

Summary. The dynamics of tree growth under pollution stress affecting leaves
and/or feeder roots have been studied using a dynamic simulation model of
basic tree processes, with parameter data for spruce. The model includes
seasonal effects and computes leaf mass in each needle age class separately,
as a result of pollution history. In the absence of pollution, the model
reproduces normal growth data for spruce. Under pollution stress, two different behavioral modes are observed: Subcritical pollution causes growth
stagnation, but allows the tree to survive, while supercritical pollution
will sooner or later lead to a sudden collapse of the tree.

DESCRIPTION OF THE SIMULATION MODEL

The tree is a living system.   The dynamic effects of leaf or root impairment on the life processes of the tree can therefore only be studied by
looking at the tree as a dynamic system composed of many interrelated and
functionally different components.   The present paper describes a dynamic
simulation model for a coniferous tree.   Further related work is reported
in Bossel 1982, Bossel/Metzler/Schäfer 1985, Metzler 1985 and Bossel 1986.

The objective of this study was to model essential processes of tree
development, and to study how impairment of essential functions (photosynthesis, feeder root renewal) would affect this development. The basic
structure of the model is shown in Fig. 1.

The possible effects of pollutants enter this system in two ways: (1) The
decrease in specific photoproductivity of the leaves reduces the available
photosynthate, and (2) a higher feeder root turnover rate drains the
assimilate pool. The quantification for the SPRUCE model is fully based on
empirical data for the different processes (e.g. specific photoproduction
as a function of light and temperature, transpiration coefficient, respiration demands, etc.). Since the model can easily be run on a small
microcomputer, the model equations are presented here as BASIC statements
(in the correct order of computation, and omitting program control statements). The results presented here were computed using the DYSYS procedure for microcomputers (Bossel 1985), but the model statements can obviously also be implemented in DYNAMO or other languages.

The model parameters, their abbreviations, dimensions, numerical values,
and respective data sources are listed in Table I. The state variables,
their abbreviations, dimensions, initial values, and respective data
sources are given in Table II. The table functions and other functions,
their abbreviations, dimensions, sources, and tables of values are presented in Table III. The auxiliary and rate equations will be documented
and explained in groups. The variables, their abbreviations and dimensions are listed in Table IV.

Leaf aging and leaf efficiency: (see the corresponding BASIC equations
below). NLAG, LFAG: The natural leaf aging rate is increased by pollution damage. NLF: The current number of needle age classes is determined by the fact that needles are shed when their efficieny limit SHED
has been reached. RPOL: The increase in leaf damage is partially offset by
an improvement due to leaf renewal. EFF: From the aging rate, the relative efficiency of each needle age class can be computed; the average
foliage efficiency is computed as the weighted average.

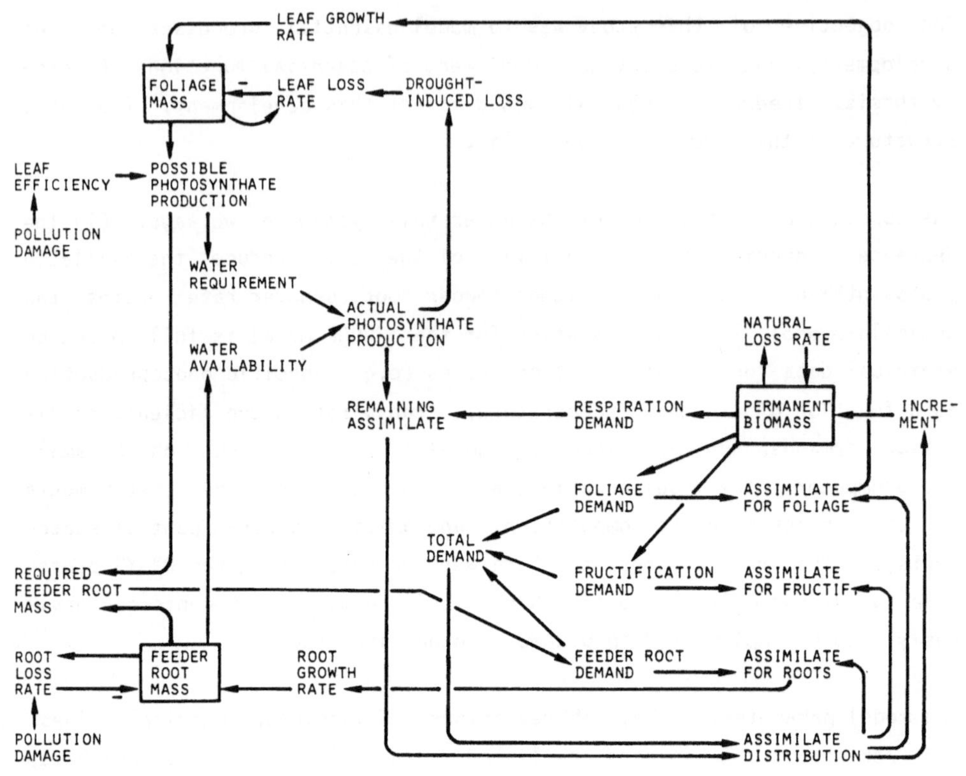

Fig. 1. **Simplified causal loop diagram for the SPRUCE model.** Photosynthate production is determined by foliage mass and transport capacity of the root system. Available assimilate is distributed according to priorities: respiration needs first, renewal needs second. The remainder, if any, is used for increment. If renewal needs are only inadequately satisfied, an accelerating process of deterioration (breakdown) begins.

```
NLAG=(1-SHED)/NDAG
LFAG=NLAG*(1+EPOL)
NLF=INT((1-SHED)/LFAG)
RPOL=LFDM-EPOL/NLF
SUM=0:FOR N=1 TO NLF:
E(N)=1-LFAG*((N-1)+TY):
SUM=SUM+E(N)*L(N): NEXT N
EFF=SUM/LEAF
```

Production of assimilate: MPRD: Assimilate production capacity is a function of leaf mass, net specific photosynthetic production, daylight period, temperature influence on photosynthesis, and relative foliage efficiency. DTRSP: This production capacity determines the required transpiration flow. TRSP: However, the actual flow is a function of feeder root mass, its transport coefficient, and soil moisture. RTRSP, PROD: If less water than required reaches the leaves, then the actual production of photosynthate is correspondingly less. RROOT, RLRT: The required feeder root mass follows from relating transport capacity to need. RTLOS: Natural feeder root decay is aggravated by root damage.

```
MPRD=LEAF*PHN*CLUX*CTEM*EFF
DTRSP=MPRD*TRSPC
TRSP=ROOT*TRANC*DROF
IF TRSP<DTRSP THEN RTRSP=TRSP/DTRSP ELSE RTRSP=1
PROD=RTRSP*MPRD
RROOT=DTRSP/TRANC
RLRT=RROOT/ROOT
RTLOS=NRTO*RTDM*ROOT
```

Respiration: RSLF, RSRT, RSST, RESP: The respiration of leaves, feeder roots, and wood biomass is a function of specific respiration coefficients and temperature. Since daylight respiration of leaves is accounted for in PHN, only dark respiration is considered in this case.

```
RSLF=LEAF*LRSP*(1-CLUX)*CTEM
RSRT=ROOT*RRSP*CTEM
RSST=BIOM*SRSP*CTEM
RESP=RSLF+RSRT+RSST
```

Distribution of assimilate if the supply is not sufficient to meet respiration needs: RAV: The relative availability of assimilate for respiration is computed by comparing the rate at which assimilate will be available to the rate at which it will be used for respiration (RESP). The rate of availability is the current production plus the consumption rate ASSI/CT of stored assimilate. LFLR, RTLR, BILR: The leaf, feeder root, and biomass losses due to insufficient respiration increase with decreasing RAV. The numerical time constants determine the speed of the deterioration due to respiration insufficiency.

```
RAV=(PROD+(ASSI/CT))/RESP
LFLR=LEAF*(1-RAV)*(52/13)
RTLR=ROOT*(1-RAV)*(52/13)
BILR=BIOM*(1-RAV)*(1/3)
RESP=RAV*RESP
```

Distribution of assimilate if the requirement for leaf and feeder root renewal cannot be fully met: NFOL: New foliage is grown only if the seasonal temperature has increased beyond TVEG. The demand for new leaves is determined by the foliage function, the normal number of needle age classes, and the amount of above-ground permanent biomass. DROOT: The assimilate demand for the growth of new feeder roots changes during the year. When the ground is too cold, no feeder root growth takes place. In the spring, before the onset of leaf growth, the demand is larger (by factor RTGF) than would be required for the mere replacement of losses; in the fall only the losses are replaced. During the vegetation period, the demand is also affected by the current relative feeder root requirement RLRT. CASS: If it turns out that the assimilates remaining after deduction of respiration needs are insufficient to meet the full demand for leaf and feeder root renewal (DGROW), then the supply reduction factor CASS<1 is applied to compute the actual amount available for both requirements (SLEAF, SROOT). SLEAF: The assimilate supply rate for leaf renewal has a time constant of 1 week (1/52 year) and is proportional to the remaining 'gap' between the foliage 'goal' (winter foliage plus new leaf demand LOLD + NFOL) and the current foliage LEAF. SINC: Since there is no assimilate surplus, there is no wood increment.

```
IF TEMP < TVEG THEN LOLD=LEAF
IF TEMP > TVEG THEN NFOL=(RFOL/NDAG)*BIOT ELSE NFOL=0
IF TEMP<TROOT THEN DROOT=0
IF TEMP>=TROOT AND TEMP<TVEG AND TY<.5 THEN DROOT=RTGF*RTLOS
IF TEMP>=TROOT AND TEMP<TVEG AND TY>.5 THEN DROOT=RTLOS
IF TEMP>TVEG THEN DROOT=RTGF*RTLOS*RLRT
DGROW=NFOL+DROOT
REST=PROD+(ASSI/CT)-RESP
CASS=REST/DGROW
IF TY<.5 THEN SLEAF=NFOL*CASS*(52/1)*(1-(LEAF/(LOLD+NFOL)))
ELSE SLEAF=0
SROOT=DROOT*CASS
SINC=0
```

Distribution of assimilate if requirements for leaf and feeder root renewal can be met: In this case, there is no reduction in SLEAF and SROOT, and the surplus assimilate is used for wood increment with a rate having a time constant of 2 weeks.

```
IF TY<.5 THEN SLEAF=NFOL*(52/1)*(1-(LEAF/(LOLD+NFOL)))
ELSE SLEAF=0
SROOT=DROOT
SINC=(REST-DGROW)*(2/52)
```

Other rates: LFDR: Leaf losses due to lack of nutrients or water appear if the flow from the feeder roots drops below 50 % of the required value. The loss rate is assumed to have a time constant of 26 weeks. LRST, LSHD: In the fall, after the end of the vegetation period, the oldest needle age class is shed at a rate having a time constant of 8 weeks.   BLDRO, NBLR: Small biomass losses occuring even under normal conditions are augmented by losses due to drought conditions.

```
RTGR=SROOT: LFGR=SLEAF
IF RTRSP<.5 THEN LFDR=LEAF*(1-RTRSP)*(52/26) ELSE LFDR=0
IF TEMP>TVEG THEN LRST=LEAF-L(NLF)
IF  TY>.5 AND TEMP<=TVEG THEN  LSHD=(LEAF-LRST)*(52/8)
ELSE LSHD=0
BLDRO=BIOM*BLDR
NBLR=BIOM*NRBL
```

The **differential equations for the state variables** are written as Euler-Cauchy integrations using BASIC.   ROOT: The feeder root mass increases with the assimilates reaching it, and decreases with turnover losses, and additional losses due to insufficient respiration activity.   ASSI: The only assimilate gains are from photosynthetic production; losses are due to respiration, leaf and feeder root renewal, and biomass growth. L(i), LEAF: The model uses a variable number of needle age groups, the current number of age classes NLF being computed from the current state of leaf damage.   At the beginning of a year, all age classes are shifted to the next higher year. The first age class is initially empty and is filled by new leaf growth during the spring (LFGR).   Losses are possible in all age classes due to deficits in nutrient and water supply (LFDR), or due to deficits in assimilate for nighttime respiration (LFLR). In addition, the oldest age class is shed in the fall (LSHD).   LEAF: The total foliage is the sum of the leaf mass contained in all age classes.   BIOM: The permanent above-ground living biomass has gains from assimilate surpluses and losses from the drying up of wood due to drought conditions, from normal losses, and from possible losses due to a deficiency in the supply of respiration assimilates.   EPOL: The pollution damage changes according to the rate computed earlier (RPOL).   These Euler-Cauchy integrations converge for timesteps DT=1/52 (1 week) or smaller.

```
ROOT=ROOT+DT*(RTGR-RTLOS-RTLR)
ASSI=ASSI+DT*(PROD-RESP-SLEAF-SROOT-SINC)
L(1)=L(1)+DT*(LFGR-(LFDR+LFLR)*L(1)/LEAF)
L(NLF)=L(NLF)-DT*(LSHD+(LFDR+LFLR)*L(NLF)/LEAF)
FOR N=2 TO (NLF-1): L(N)=L(N)-DT*((LFDR+LFLR)*L(N)/LEAF): NEXT N
LEAF=0: FOR N=1 TO NLF: LEAF=LEAF+L(N): NEXT N
BIOM=BIOM+DT*(SINC-BLDRO-NBLR-BILR)
EPOL=EPOL+DT*RPOL
```

SIMULATION RESULTS

The simulation model produces three behavioral modes: (1) normal growth
under subcritical pollution stress, (2) stagnation for somewhat higher
(subcritical) stress, and (3) eventual breakdown of the tree under supercritical conditions. The reasons for these distinct behavioral modes can
be found by identifying the dominant system loops in each case (Fig. 2).

Due to the complexity of the SPRUCE model (especially the table functions
and other nonlinearities), analytic expressions for the stagnation and
breakdown limits cannot be easily formulated. However, the numerical
computation of these limits is straightforward. Fig. 3 shows the results
from 250 simulations for young spruce (initial permanent biomass BIOMO =
50 t ODM / ha). The diagram indicates whether a given combination of leaf
and feeder root damage will (1) allow growth (small dots), or (2) result
in stagnation (medium dots), or (3) lead to tree breakdown (large dots).

An important conclusion from these studies is the observation that younger
trees have a much better tolerance to pollution damage than older trees.
In fact, pollution levels leading to certain breakdown of older trees may
still allow some growth in younger trees.

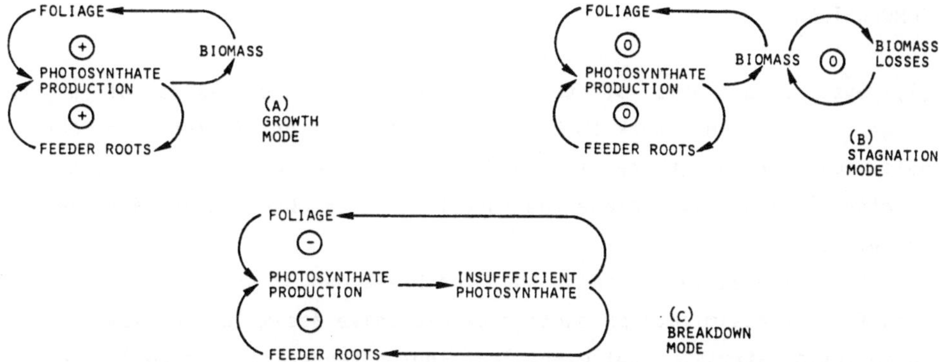

Fig. 2. **Dominant system loops determining the three modes of** behavior. (A) In the growth mode, surplus photosynthate leads to increase in biomass, foliage, and feeder roots, and hence further enhancement of photosynthetic production. (B) In the stagnation mode, photosynthate production just suffices to keep up with natural losses of biomass, leaves, and feeder roots. (C) In the breakdown mode, an insufficient supply of photosynthate causes incomplete replacement of leaves and feeder roots, leading to further accelerated reduction of photosynthetic production, and of leaf and feeder root mass.

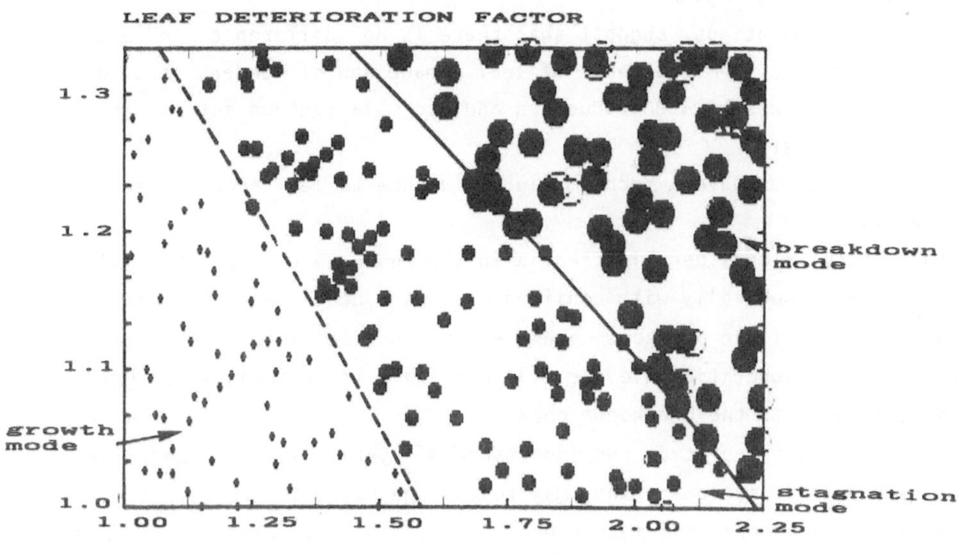

Fig. 3. **Behavioral mode diagram for young spruce,** obtained from 250 simulations using random values for leaf deterioration factor LFDF and root turnover factor RTOF. (1) Growth mode: small dots; (2) Stagnation mode: intermediate dots; (3) Breakdown mode: large dots. Initial conditions: BIOM=50, ASSI=2.5, ROOT=7, LEAF=7, NDAG=7. Simulation over 20 years using the SPRUCE model. The diagram clearly shows the existence of three distinct behavioral regions.

CONCLUSIONS

(1) Systems analysis of the structure and major processes of a tree suggests forest die-back to be a 'natural' reaction of trees to chronic pollution stress: there is probably no need to search for a mysterious 'factor X' which would cause the forest die-back observed in many parts of the world.

(2) The simulations suggest that the breakdown of trees is caused by insufficient assimilate production or excessive assimilate consumption due to pollution stress. Deficient leaf and/or feeder root renewal leads to increasingly rapid deterioration.

(3) The breakdown mode will only occur if the assimilate deficieny reaches a critical level. This may occur as the result of chronic pollution stress over many years.

(4) It therefore makes sense to speak of 'subcritical' stress (no break-down, but reduced growth) and 'supercritical' stress (eventual breakdown).

(5) In the subcritical regime, the tree still looks 'healthy', even though growth may be absent. Two distinct modes - growth and stagnation - are possible in this regime.

(6) The simulations suggest that there is no difference in principle between the systemic effects of leaf damage and of feeder root damage: both reduce assimilate production and accumulation and act on the same life processes.

(7) Older trees have a much smaller tolerance to leaf and/or feeder root damage.

(8) There is considerable inertia in the response of trees to pollution reduction, especially with conifers with many needle age classes. Also, the dynamics of the breakdown phase are self-sustaining. Hence pollution reduction to subcritical levels will only save those trees which have not yet entered into the breakdown phase.

(9) The results point to the importance of systems analysis and modelling of ecological systems: they may reveal the systemic causes of strange behavior which empirical research of isolated details would not be able to clarify.

(10) While the model presented here cannot be used for precise forecasts of forest development, it can be expected that simulation models improved with respect to structure and data could be used for regional forecasts of forest growth and die-back as the result of a given pollution spectrum.

REFERENCES

Bossel, H., 1985. Umweltdynamik - 30 Programme fr kybernetische Umwelt-
erfahrungen auf BASIC-Rechnern. Te-Wi Verlag, Munich.

Bossel, H. et al., 1982. Dynamik von Waldökosystemen - Mathematisches
Modell und Computersimulation. Interdisziplinäre Arbeitsgruppe Mathemati-
sierung (IAGM), University of Kassel.

Bossel, H., Metzler, W., Schäfer, H. (Editors), 1985. Dynamik des Wald-
sterbens - Mathematisches Modell und Computersimulation. Springer Verlag,
Berlin/Heidelberg/New York/Tokyo.

Eidmann, F.E., Schwenke, H.J., 1967. Beiträge zur Stoffproduktion, Transpiration und Wurzelatmung einiger wichtiger Baumarten. Forstwissen-
schaftliche Forschungen, Heft 23, Parey, Hamburg/Berlin.

Finck, S., 1982. Pflanzenernährung in Stichworten. Hirt, Kiel.

Hunt, R., 1982. Plant growth curves - The functional approach to plant
growth analysis. E. Arnold, London.

Larcher, W., 1980. Oekologie der Pflanzen auf physiologischer Grundlage.
E. Ulmer, Stuttgart.

Mayer, H., 1977. Waldbau auf soziologisch-ökologischer Grundlage. G.
Fischer Verlag, Stuttgart/New York.

Metzler, W., 1985. Dynamical systems modelling and simulation of air
polluted forest ecosystems. Proceedings, 5th International Conference on
Mathematical Modelling, Berkeley, July 1985.

Mitscherlich, G., 1971. Wald, Wachstum und Umwelt. Vol. 1: Form und
Wachstum von Baum und Bestand. Sauerländer, Frankfurt/M.

McClaugherty, C.A., Aber, J.D., 1982. The role of fine roots in the
organic matter and nitrogen budgets of two forested ecosystems. Ecology,
63(5), p. 1481-1490.

Rehfuess, K.E., 1981. Waldböden. Parey, Hamburg/Berlin.

Röhrig, E., 1966. Die Wurzelentwicklung der Waldbäume in Abhängigkeit von
den ökologischen Verhältnissen. Forstarchiv, 37. Jahrgang, Heft 10, p.
217-229 and Heft 11, p. 237-249.

Schmidt-Vogt, H., 1977. Die Fichte. Parey, Hamburg/Berlin 1977.

Sharpe, P.J.H., 1983. Responses of photosynthesis and dark respiration to
temperature. Annals of Botany 52, pp. 325-348.

TABLE I - PARAMETERS FOR THE SPRUCE MODEL

CT   = 2/52 (a)
      Assimilate consumption time
LFDF >= 1 (-)
      Leaf deterioration factor. LFDM = LFDF-1.
LRSP = 1.5 ((kg assimilate/a)/kg leaf ODM)
      Specific dark respiration of leaves
      (converted from data in Larcher 1980, p. 138)
LUX   = 250 (W/m2)
      Average insolation during daylight hours
      (value not critical; light saturation occurs at 100 W/m2;
      Larcher p. 139-141)
NDAG = 8 (a)
      Normal needle shedding age
      (Schuett 1983, p.34)
NRBL = 0.01 (1/a)
      Normal wood biomass loss (relative)
      (Larcher 1980, p. 186, modified)
NRTO = 1 (1/a)
      Normal feeder root decay rate
      (McClaugherty/Aber 1982, p. 1481)
RRSP = 3 ((kg assimilate/a)/kg f.root ODM)
      Specific respiration of feeder roots
      (after Eidmann/Schwenke 1967, Roehrig 1966, McClaugherty 1982)
RTGF = 2 (-)
      Feeder root growth factor
      (McClaugherty/Aber 1982, modified)
RTOF >= 1 (-)
      Feeder root turnover factor. RTDM = RTOF.
SHED = 0.2 (-)
      Relative leaf efficiency for shedding
      (Larcher 1980, p. 137)SRSP = 0.05 ((kg assimilate/a)/kg biomass ODM)
      Specific respiration of stem and coarse roots
      (after Larcher 1980, p. 134, p. 186)
SWP   = 0 (bar)
      Soil water potential
      (Assumption: water saturation)
TRANC = 2000 ((kg water/a)/kg f.root ODM)
      Transport coefficient
      (recomputed from data in Eidmann/Schwenke 1967)
TROOT = 6 (degrees C)
      Average soil temperature marking begin and end of feeder
      root growth (from Mitscherlich p. 54, and temperature data
      for Kassel)
TRSPC = 241 (kg water/kg assimilate)
      Transpiration coefficient
      (Eidmann/Schwenke 1967, p. 17)
TVEG = 10 (degrees C)
      Average air temperature marking begin and end of vegetation
      period (Larcher 1980, p. 87; Mitscherlich 1971, p. 54)

TABLE II - STATE VARIABLES AND INITIAL VALUES FOR THE SPRUCE MODEL

ASSI = 6.25 (t ODM/ha)
     Stock of assimilate/photosynthate
     (inventory for Jan.1)
BIOM = 250 (t ODM/ha)
     Permanent biomass above and below ground
EPOL = 0 (1/a)
     Air pollution effect on leaf efficiency
     (no damage initially)
L(1)...L(NLF) = 12.5/8 (t ODM/ha)
     Leaf mass per needle age class
     (corresponding to a foliage mass of LEAF = 12.5 for NLF = 8)
ROOT = 12.5 (t ODM/ha)
     Feeder root mass

TABLE III - TABLE FUNCTIONS AND OTHER FUNCTIONS FOR THE SPRUCE MODEL

BLDR (1/a)
     Biomass loss due to drought
     (after Larcher 1980, p. 343, and Finck 1982, p. 80)

SWP	0	10	15	30
BLDR	0	0	1	1

CLUX = (1/24)*(12+4*sin(6.28*TY-1.57))
     Daylight hour fraction
     (approximate relationship for the latitude of Kassel)
CTEM (-)
     Temperature influence on photosynthesis and dark respiration
     (after Larcher 1980, p. 150, p. 152, and Sharpe 1983, p. 337)

TEMP	-5	5	20	30
CTEM	0	.2	1	1

DROF (-)
     Drought factor
     (after Schmidt/Vogt 1977, p. 413, and Larcher 1980, p. 292)

SWP	0	2	15	20	30
DROF	1	1	.15	.1	0

PHN ((kg assimilate/a)/kg leaf ODM)
     Net photosynthesis
     (after Larcher 1980, p. 138, p. 141)

LUX	0	5	100	1000
PHN	-1.5	0	24	24

RFOL (kg leaf mass ODM/kg biomass ODM)
     Relative foliage function
     (after Bossel et al. 1982, p. 106; Mayer 1977, p. 40;
     Mitscherlich 1971, p. 30; Larcher 1980, p. 168; see also
     discussion of the Gompertz 1825 function in Hunt 1982)

LBIOT = ln(ln BIOT)	0.0	0.9	2.08
RFOL	0.7	0.18	0.0

     where BIOT = BIOM+LEAF+ROOT
TEMP (degrees C)
     Air temperature function (monthly average)
     (long term weather data for Kassel

TY	.0	.058	.135	.212	.308	.385	.462	.558	.635	.731	.808	.885	.961	1.
TEMP	.7	0.0	0.8	4.6	8.8	13.2	16.4	17.8	17.3	14.2	9.1	4.9	1.5	.7

TABLE IV - AUXILIARY VARIABLES AND RATES FOR THE SPRUCE MODEL

BILR   permanent biomass loss rate to due respiration assimilate
       deficiency ((kg ODM/a)/ha)
BIOT   total biomass = BIOM+LEAF+ROOT (t ODM/ha)
BLDRO biomass loss due to drought ((kg ODM/a)/ha)
CASS   assimilate supply reduction factor (-)
DGROW assimilate demand for feeder roots and leaf growth ((kg/a)/ha)
DROOT assimilate demand for feeder root formation ((kg/a)/ha)
DTRSP required transpiration flow ((kg water/a)/ha)
E(n)   relative photosynthetic efficiency of needle age class n (-)
EFF   average relative foliage efficiency (-)
LEAF   total foliage mass (kg ODM/ha)
LFAG   actual leaf aging rate (1/a)
LFDR   leaf shedding rate due to transpiration deficit ((kg ODM/a)/ha)
LFGR   leaf growth rate ((kg ODM/a)/ha)
LFLR   leaf loss rate due to respiration assimilate deficiency ((kg
LOLD   foliage mass before new leaf growth (kg ODM/ha)
LRST   foliage mass remaining after shedding (kg ODM/ha)
LSHD   shedding rate for the oldest needle age class ((kg ODM/a)/ha)
MPRD   photoproduction capacity ((kg assimilate/a)/ha)
NBLR   normal biomass loss rate ((kg ODM/a)/ha)
NFOL   normal new leaf mass (kg ODM/ha)
NLAG   natural leaf aging rate (1/a)
NLF   number of needle age classes (-)
PROD   actual production of photosynthate ((kg assimilate/a)/ha)
RAV   relative assimilate availability (-)
RESP   total respiration ((kg assimilate/a)/ha)
REST   assimilate remaining for leaf, feeder root, and wood growth
       ((kg/a)/ha)
RLRT   relative required feeder root mass (-)
RPOL   net leaf damage rate (1/a)
RROOT required feeder root mass (kg f.root ODM/ha)
RSLF   dark respiration of leaves ((kg assimilate/a)/ha)
RSRT   respiration of feeder roots ((kg assimilate/a)/ha)
RSST   respiration of stem and coarse roots ((kg assimilate/a)/ha)
RTGR   feeder root growth rate ((kg ODM/a)/ha)
RTLOS feeder root decay rate ((kg ODM/a)/ha)
RTLR   feeder root loss rate due to respiration assimilate defi-
       ciency ((kg ODM/a)/ha)
RTRSP relative transpiration flow (-)
SINC   assimilate supply for wood growth ((kg/a)/ha)
SLEAF assimilate supply for leaf growth ((kg/a)/ha)
SROOT assimilate supply for feeder root growth ((kg/a)/ha)
T   time (a)
TRSP   actual transpiration flow ((kg water/a)/ha)
       ODM/a)/ha)
TY   vegetation time (a) (Jan.1 = 0, Dec. 31 = 1)

a   year
ODM   organic dry matter

Selected Papers

Applying system analysis techniques for biomedical research represents a most challenging and difficult problem. The proceedings of the 2nd Ebernburger Working Conference are structured with this in mind: they illustrate several important aspects of system analysis considered within a biomedical system science context:

Methods

Biological and Physiological Systems

Methods

Fit, Fitter, The Fittest
Methods for Modeling and Identification of Dynamical Systems
H.G. Bock, J.P. Schlöder

Strange Limit Cycles, Chaos, and Invariant Measure for a simple
Differential-Delay Equation arising from Biology
U. an der Heiden

The use of Computer Simulation to Evaluate the Testability of a
new Fitness Concept
W. Gabriel

Texture Analysis using Random Field Models exemplified on Ultrasonic Images of the Liver
U. Ranft

Smooth Descriptive Modelling of Multifactorial Systems Responses
G.H. Klein

Zur Beschreibung offener Thermodynamischer Prozesse durch
Bindungsdiagramme
A. Schöne

Fit, Fitter, The Fittest – Methods for Modeling and Identification of Dynamical Systems

H. G. Bock and J. P. Schlöder

Institut für Angewandte Mathematik und Sonderforschungsbereich 72, Universität Bonn, Wegelerstr. 10, D-5300 Bonn 1, Fed. Rep. of Germany.

THE USE OF IMPLICIT MODELS FOR DYNAMICAL PROCESSES CHARACTERIZATION

Modeling a dynamical process is a difficult and involved task. To obtain a quantitatively correct model, which reproduces the real behaviour of a dynamical process under a variety of different conditions, requires the conduction and evaluation of a corresponding variety of experiments, which exhibit the characteristic properties of the dynamic process: the dynamic behaviour, possible steady states, unstable steady states, periodic solutions, Hopf-bifurcation points and so on.

The description of these phenomena is naturally implicit: the basic model is typically a differential-equation system in non-stationary or stationary form, and situations like unstable steady states or periodicity of solutions are characterized by implicit features of the model, as is shown in table 1 for a number of typical cases in ODE-modeling.

<u>**Table 1:**</u> Characterization of specific situations of a dynamical process by ODE-models (k : Index of experiment)

Experimental situation	Characterization	Observations η
general dynamical behaviour	(1) $\dot{y}^k - f(y^k, p\,\|q^k) = 0$ or $\quad F(y^k, \dot{y}^k, p\,\|q^k) = 0$	$\eta_{ij}^k = g_j(t_i, y^k(t_i), p\,\|q^k) + \epsilon_{ij}^k$
periodic solutions	(1) $\quad$ and (2) $y^k(t) = y^k(t + T^k)$	as for (1) including $\quad \eta_T^k = T^k + \epsilon_T^k$.
steady state	(3) $f(y^k, p\,\|\,q^k) = 0$	$\eta_{ij}^k = h_{ij}(y^k, p\,\|q^k) + \epsilon_{ij}^k$
unstable steady state	(3) $\quad$ and (4) $\frac{\partial f}{\partial y}(y^k, p\|q^k) \cdot h^k = 0$ $\quad h^{k\top}h^k - 1 = 0$	as for (3)
Hopf-bifurcation	(3) $\quad$ and (5) $\frac{\partial f}{\partial y}(y^k, p\,\|q^k) \cdot h_1^k + \alpha h_2^k = 0$ $\frac{\partial f}{\partial y}(y^k, p\,\|q^k) \cdot h_2^k - \alpha h_1^k = 0$ $+\ 2\ reg.\ cond.$	as for (3)

The observations or measurements η are considered to be functions (g) of the "true values" of the *state variables* y and of the *parameters* p and q^k of the model (assumed to be unknown) which are perturbed by noise ϵ. For ease of presentation we will assume the standard statistical case of independent normally distributed noise with known variances σ_η^2 and zero mean $(N(0, (\sigma_{ij}^k)^2))$.

The quantities q^k are parameters that describe the specific situation of experiment No. k, they are *independent variables* in the statistical sence (like the observation times $\{t_j^k\}$, too). They may also be parametrizations of certain input functions of the dynamical system.

<u>**Remark:**</u> Quantities such as y, $\{t_j\}$, q, ϵ, g, T, h may differ from experiment to experiment, whereas the system parameters p are assumed to be invariant.

Note, that the distinction between q and p is sometimes a structural rather than a physical one.

INTRINSIC DIFFICULTIES OF DYNAMICAL SYSTEMS

Realistic modeling of dynamical systems "in the large" (i.e. not only in a neighbourhood of a steady state) usually requires *nonlinear differential equation systems*. The differential equation solution of system (1) can be *stiff* or *unstable*, and these properties may vary with time or changes of system parameters.

At least for certain values of the parameters a solution may not even exist, or may be hard to obtain numerically.

Closely related to this is the fact, that the *implicit characterizations* of the model as given in table 1 can in general not be reformulated in an explicit form. Even if in principle possible, the analytical effort may be prohibitive. From the numerical analysts point of view, it also appears to be inadequate because the explicit a priori inversion of an implicit equation usually deteriorates the *numerical* well-posedness of a problem. It is therefore advisable if not indispensable to treat such cases *directly*. The fitness of solution approaches depends very much on whether they fit these requirements or not, otherwise, they may give you a fit.

VALIDATION LEADS TO LARGE STRUCTURED PARAMETER ESTIMATION PROBLEMS

Large scale problems in parameter estimation arise from a number of different reasons:

(i)   Validation of a model under different conditions requires the determination of system parameters from data obtained under different conditions. Therefore, measurements from *large numbers of different experiments* may be fitted to simultaneously.

(ii)   The differential equation system (1), (3) may be of high dimension. Ecosystems can easily reach a few dozens of state variables, chemical reaction systems can reach a few hundred. Dimensions can go beyond thousands as soon as (1) is a discretization of partial differential equations. Even in genuine ODE-problems, the dimension of a parameter estimation problem can reach a few ten thousands. Note, however, the principle question of *identifiability* of large models.

An Unconventional Problem Format

If measurements from many different experiments are used, parameter identification problems of the following structure occur:

$$\text{minimize} \quad \sum_{k=1}^{N} \|u_k(x_k, x_g)\|_2^2$$

(6)

N : No. of experiments

x_k : local variables (depending on experiment k)

x_g : global variables

subject to nonlinear equality constraints

(7)
$$v_k(x_k, x_g) = 0$$

and possible inequality constraints

$$(8) \qquad\qquad w_k(x_k,\ x_g) \geq 0.$$

Equality constraints represent additional model descriptions, inequality constraints serve as a means to consider certain a priori information about the model, such as positivity bounds. The minimization of the cost function (6) yields a maximum likelihood estimate of the unknown parameters, if we set (case 3, e.g.)

$$u_k =: \left\{ (\eta_i^k - h(y_i^k, p\, |q^k))/\sigma_{\eta_i^k} \right\}_{i=1}^{n_k}$$

GENERALIZED GAUSS-NEWTON-METHODS FOR LARGE NONLINEAR CONSTRAINED LEAST SQUARES PROBLEMS

Both theory and practice of GGN-methods for large scale constrained least squares were developed in references [1-5, 8, 9] The *basic procedure* is to improve a certain iterate $z^j := (x_1^j,\ ...,\ x_N^j,\ x_g^j)$ by searching along a given direction Δz_j

$$(9) \qquad\qquad z^{j+1} = z^j + t_j \Delta z^j, \quad 0 < t_j \leq 1$$

which is given as a solution of the *linearized system*

$$
\begin{aligned}
\min &= \sum_{k=1}^{N} \| u_k + \frac{\partial u_k}{\partial x_k} \cdot \Delta x_k + \frac{\partial u_k}{\partial x_g} \Delta x_g \|_2^2 \\
\text{s.t.} \quad & v_k + \frac{\partial v_k}{\partial x_k} \Delta x_k + \frac{\partial v_k}{\partial x_g} \Delta x_g = 0 \\
& w_k + \frac{\partial w_k}{\partial x_k} \Delta x_k + \frac{\partial w_k}{\partial x_g} \Delta x_g \geq 0.
\end{aligned}
$$

(10)

Here, $(u,v,w)_k$ and $\frac{\partial(u,v,w)_k}{\partial(x_k,\ x_g)}$ are evaluated at $\left(x_k^j,\ x_g^j \right)$.

Procedures of this type have proven very successful. Their basic property is that they exploit the *sensitivity matrices of the model* $(\partial(uvw)/\partial(x_k x_g))$ in every iterative step.

They outperform low order methods such as trial & error, random search or systematic search. The latter do not evaluate sensitivity matrices, and are thus not suitable even for mildly illconditioned problems. As soon as one unknown is better determined than another long and narrow "banana valleys" occur, which cause this kind of procedure to terminate very soon. Moreover, it is a professional error to try to solve implicit models this way.

It can easily be seen, that (10) actually leads to an "almost second" order expansion of the cost function

$$
\left\| u_k + \frac{\partial u_k}{\partial x_k} \Delta x_k + \frac{\partial u_k}{\partial x_g} \Delta x_g \right\|_2^2
$$

$$
= \| u_k \|_2^2 + 2 u_k^{\top} \left(\frac{\partial u_k}{\partial x_k}, \frac{\partial u_k}{\partial x_g} \right) \begin{pmatrix} \Delta x_k \\ \Delta x_g \end{pmatrix}
$$

(11)
$$
+ \begin{pmatrix} \Delta x_k \\ \Delta x_g \end{pmatrix}^{\top} \underbrace{\begin{pmatrix} \frac{\partial u_k}{\partial x_k}^{\top} \frac{\partial u_k}{\partial x_k} & \frac{\partial u_k}{\partial x_k}^{\top} \frac{\partial u_k}{\partial x_g} \\ \frac{\partial u_k}{\partial x_g}^{\top} \frac{\partial u_k}{\partial x_k} & \frac{\partial u_k}{\partial x_g}^{\top} \frac{\partial u_k}{\partial x_g} \end{pmatrix}}_{*} \begin{pmatrix} \Delta x_k \\ \Delta x_g \end{pmatrix}
$$

which leaves out a term of the Hessian of the Lagrange-function the expected value of which is zero at the true solution. Consequently, the convergence properties are in general better than all first order methods of gradient type, which make no use of the second order information, present in (*). Thus, GGN variants appear to be the fittest of all fitting methods.

Convergence properties

Can be briefly summarized as follows:

(i)   The local convergence is linear with a convergence rate that improves, the better the compatibility of model & data ("effective second order").

(ii)   The method is locally attracted only by "statistically stable" minima, which rules out "far away" local minima.

(iii)   The method is theoretically globally convergent, i.e. in practice robust for bad initial guesses.

The use of the sensitivity matrices in every step as in (11) allows the explicit discrimination between ill- or well-conditioned parameters – a banana valley approximation is used. Moreover, special step strategies are used which make sure that the optimization procedure is not trapped in the banana valley but is allowed to march fast along the rim.

GENERATION AND SOLUTION OF GGN
LARGE STRUCTURED LINEAR SYSTEMS

The most involved task – both from computing time and realization aspect – is the *generation* and the *solution* of the linear problems corresponding to (10).

Figure 1 shows the block structures of the Jacobians/sensitivity matrices associated with validation problem.

Quite clearly, any method that does not take into account the structure of the problem is not competitive even for small scale problems. Therefore, special techniques for the treatment of practical PI problems were developed in [8,9], which are specifically designed for validation structures.

Fig. 1. Block structure of multiple experiment Jacobian E_{Lk}, E_{Gk} Jacobian of constraints, U_{Lk}, U_{Gk} Jacobian of least squares conditions

$$(12) \qquad \begin{pmatrix} E_{L1} & 0 & 0 & 0 & E_{G1} \\ U_{L1} & & & & U_{G1} \\ 0 & E_{L2} & 0 & 0 & E_{G2} \\ & U_{L2} & & & U_{G2} \\ 0 & 0 & E_{L3} & 0 & E_{G3} \\ & & U_{L3} & & U_{G3} \\ & & & \ddots & \vdots \\ 0 & 0 & 0 & E_{LN} & U_{GN} \\ & & & U_{LN} & U_{GN} \end{pmatrix}$$

The individual experiment blocks can have considerably different format, e.g. if they are of different types according to table 1. Of course, they may exhibit special substructures themselves, that have to be taken into account in large scale problems.

BVP METHODS FOR THE DYNAMIC CASE

In the non-stationary case 1 (or 2), stable and efficient ways of treatment by *multiple shooting* or *collocation* have been introduced in recent years (see [1-5,8,9]) which may be briefly recalled.

The time interval under consideration is divided into subintervals

$$(13) \qquad t_o < t_1 < \ldots t_{m-1} < t_m,$$

where the *nodes* $\{t_j\}$ are e.g. a subset of the observation points, and the values of the state variables $s_j \doteq y(t_j)$ at the nodes are introduced as additional variables.

We then solve the ODE-system on each of the m subintervals (in case of multiple shooting), starting from an initial estimate s_j^o to obtain a *discontinuous* function

$$(14) \qquad y(t;\, s_j,\, p) \quad t_j \leq t \leq t_{j+1}.$$

The satisfaction of the ODE-system on the whole interval is then achieved by requiring the continuity at the nodes

$$(15) \qquad y(t_{j+1};\, s_j,\, p) = s_{j+1}$$

as an *implicit* side condition which determines the extra variables $\{s_j\}$.

The multiple shooting (or more general: the boundary value problem approach) has two distinct advantages over simpler initial value techniques, which integrate system (1) over the whole interval.

(i)   The method is numerically stable, and it can be used to solve PI-problems for ODE-systems with highly unstable components.

(ii)   Even if this is not neccessary – as in stiff stable systems – the method has superior efficieny properties, because the additional variables s_j allow to bring in information about the state variables into the problem formulation, which damps the influence of bad estimates on the convergence behaviour of the optimization procedure (see examples in [4,5,8]).

The linearized system for a dynamic experiment takes the form

$$
(16) \qquad
\begin{aligned}
U_{Lk} &= \left\{ \begin{bmatrix} D_1^1 & D_1^2 & . & . & . & D_1^q \end{bmatrix} \right. \\
E_{Lk} &= \left\{ \begin{bmatrix}
D_2^1 & D_2^2 & . & . & . & D_2^q \\
G_1 & -I & & & & G_1^q \\
& G_2 & -I & & & G_2^q \\
& & \ddots & \ddots & & \vdots \\
& & & G_{m-1} & -I & G_{m-1}^q
\end{bmatrix} \right.
\end{aligned}
\qquad
\begin{aligned}
U_{Gk} &= \left\{ \begin{bmatrix} D_1^p \end{bmatrix} \right. \\
E_{Gk} &= \left\{ \begin{bmatrix}
D_2^p \\ G_1^p \\ G_2^p \\ \vdots \\ G_m^p
\end{bmatrix} \right.
\end{aligned}
$$

Fig. 2: Zoom on structure of a dynamic case block in Fig. 1 for multiple experiment systems

The dynamic case reveals further block sparse structure, which depends basically on the discretization scheme. For collocation and finite difference similar structures appear – see Bär [1] and [3].

GENERATION AND DECOMPOSITION: DYNAMIC CASE

The main computational effort definitively arises in the generation and decomposition of the system shown in Fig. 2.

(1)   The significant work in the generation of the derivative matrix comes from differentation of the initial value problem solutions $y(t; s_i, p)$ in eq. (14). To avoid programming errors and save computing time, these derivatives are automatically generated by *"internal numerical differentiation"* ([2],[4], [8]).

(2)   Schlöder [8] introduces a unique way of *simultaneously* generating and decomposing system (16), using all implicit or explicit restrictions in the side conditions to reduce the degrees of freedom in the system. The speed-up is amazing especially for very large systems with a few genuine degrees of freedom. See [8] for a detailed derivation, also [4,5] for an outline.

(3)   Specific decomposition techniques with superior stability properties were introduced in [4,5]. They allow for the treatment of extremely unstable systems, and can also be used for a detection of potentially incorrectly posed side conditions.

STATISTICAL SENSITIVITY ANALYSIS

The solution of parameter estimation problems is useful for model validation purposes only if a statistical system analysis is provided. Algorithms for computation of variance-covariance statistics – based on generalized inverse theory – for the estimation problem in the unconventional format as given above were derived in [2,8,9]. More recently, optimization procedures for minimization of confidence regions and intervals, resp., were designed that are based on these computational techniques for the covariance matrix, or its diagonal elements. Such algorithms will play a significant role in the design of optimal experiments.

VALIDATION OF THE OSCILLATING BELOUSOV-ZHABOTINSKII REACTION FROM BISTABILITY POINTS

The oscillations of the Belousov-Zhabotinskii-reaction are one of the most fascinating phenomena in physical chemistry. From the many models proposed to explain its nature, we consider here the Noyes-Field-Thompson mechanism [7], which describes the anorganic part of the BZR in a CSTR.

$$
\begin{aligned}
HBrO_2 \quad & \dot{A} = p_1 BCH^2 - p_{-1} AD - p_2 ACH + p_{-2} D^2 - p_4 ABH + p_{-4} F^2 + p_5 FGH \\
& \quad - p_{-5} AI - 2p_7 A^2 + 2p_{-7} BDH - kA \\[4pt]
BrO_3^- \quad & \dot{B} = -p_1 BCH^2 + p_{-1} AD - p_4 ABH + p_{-4} F^2 + p_6 FI - p_{-6} BGH^2 + p_7 A^2 \\
& \quad - p_{-7} BDH - k(B - B_o) \\[4pt]
Br^- \quad & \dot{C} = -p_{-1} BCH^2 + p_{-1} AD - p_2 ACH + p_{-2} D^2 - p_3 CDH + p_{-3} E - k(C - C_o) \\[4pt]
HOBr \quad & \dot{D} = p_1 BCH^2 - p_{-1} AD + 2p_2 ACH - 2p_{-2} D^2 - p_3 CDH + p_{-3} E + p_7 A^2 \\
& \quad - p_{-7} BDH - kD \\[4pt]
Br_2 \quad & \dot{E} = p_3 CDH - p_{-3} E - kE \\[4pt]
BrO_2 \quad & \dot{F} = 2p_4 ABH - 2p_{-4} F^2 - p_5 FGH + p_{-5} AI - p_6 FI + p_{-6} BGH^2 - kF \\[4pt]
Ce^{3+} \quad & \dot{G} = -p_5 FGH + p_{-5} AI + p_6 FI - p_{-6} BGH^2 - (G - G_o) \\[4pt]
H^+ \quad & \dot{H} = -2p_1 BCH^2 + 2p_{-1} AD - p_2 ACH + p_{-2} D^2 - p_3 CDH + p_{-3} E - p_4 ABH \\
& \quad + p_{-4} F^2 - p_5 FGH + p_{-5} AI + 2p_6 FI - 2p_{-6} BGH^2 + p_7 A^2 \\
& \quad - p_{-7} BDH - k(H - H_o) \\[4pt]
Ce^{4+} \quad & \dot{I} = p_5 FGH - p_{-5} AI - p_6 FI + p_{-6} BGH^2 - kI
\end{aligned}
\tag{17}
$$

This ODE system contains 14 unknown rate constants to be determined. Geiseler and Bar-Eli [6] have published data on the approximate location (in terms of the five experimental parameters retention time and 4 inflow concentrations) of 44 unstable steady states. A formulation of the corresponding parameteridentification problem comprises a system with 44 multiple experiments of case (4) with 14 global parameters $p_{\pm 1}, \ldots, p_{\pm 7}$, 9 local concentrations $\{y_i^k\}_{i=1}^9$, 9 local auxiliary values $\{h_i^k\}_{i=1}^9$ and 2 local experimental parameters (which are known subject to error only), in total 894 variables. On the other hand there are 2 least squares and 19 equality conditions per experiment and in addition, 2 global thermodynamical constraints.

$$
p_4 \cdot p_5 \cdot p_6 = p_{-4} \cdot p_{-5} \cdot p_{-6}
$$

$$
p_1 \cdot p_{-2} \cdot p_7 = p_{-1} \cdot p_2 \cdot p_{-7}
$$

which amounts to 926 conditions, in addition to which 498 positivity bounds on concentrations, experimental parameters and rate constants have to be taken into account.

To the experience of the authors, the problem is both *extremly* nonlinear and ill-conditioned.

Numerical results for this problem are given in figures 3-9. They show a good visual agreement of data and model. However, the root mean square error is still approximately 15% in contrast to 5% measurement errors – according to the experimentators information [6].

The misfits appear to be caused by errors in the H^+- part of the model, as can be seen in figure 6 and 7. This indicates that still more modeling is required. On the other hand, it turns out that the standard errors of the rate constants are still rather large, which means that more experimental data should be included – particularly of dynamic experiments – in order to reduce the confidence regions.

For selection of these experiments, the optimal experiment design procedures mentioned above could prove very useful.

Fig. 3-9 Solution (o measured data, + computed result)

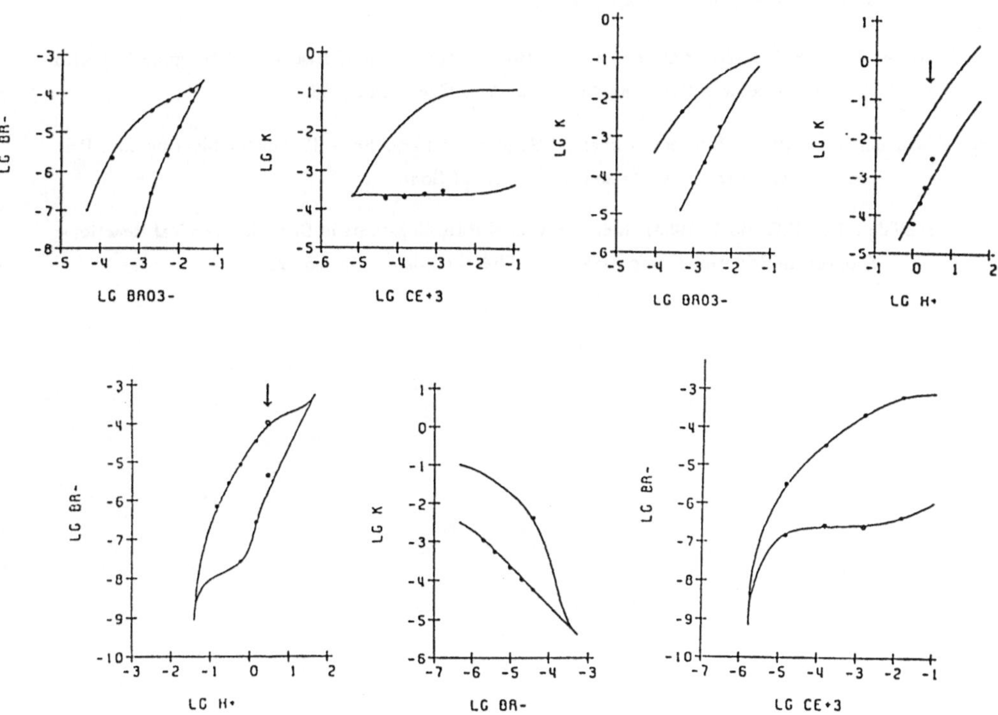

REFERENCES

[1]   Bär, V. (1983). A Collocation Method for Numerical Solution of General Multipoint Boundary
        Value Problems (in German), Diplom Thesis, University of Bonn.

[2]   Bock, H.G. (1981). Numerical Treatment of Inverse Problems in Chemical Reaction Kinetics,
        Modelling of Chemical Reaction Systems, Springer Series in Chemical Physics 18, pp. 102-
        125.

[3]   Bock, H.G. (1983). Recent Advances in Parameter Identification Techniques for O.D.E.-Systems,
        Progress in Scientific Computing 2, Birkhäuser Boston, pp. 95-121.

[4]   Bock, H.G. (1985). Boundary Value Problem Methods for Parameter Identification in Systems of
        Ordinary Differential Equations (in German), to appear in Bonner Math. Schriften.

[5]   Bock, H.G., J.P. Schlöder (1986). Recent Progress in the Development of Algorithms and Software
        for Parameter Estimation in Large Chemical Reaction Systems, Proc. of IFAC Workshop
        on Automatic Control in Petroleum, Petrochemical and Desalination Industries, Kuwait.

[6]   Geiseler, W., K. Bar-Eli (1981). Bistability of Oxidation of Cerous Ions by Bromate in a Stirred
        Flow Reactor, J. Phys. Chem., 85, 908.

[7]   Noyes, R.M., R.J. Field, R.C. Thompson (1971). Mechanism of Reaction of Bromine (V) with
        Weak One-Electron Reducing Agents, J. Am. Chem. Soc., 93, 7315

[8]   Schlöder, J.P. (1986). Efficient Numerical Solution of Large Scale Parameter Identification Pro-
        blems (in German), Dissertation, University of Bonn.

[9]   Schlöder, J.P., H.G. Bock (1983). Identification of Rate Constants in Bistable Chemical Reactions,
        Progress in Scientific Computing 2, Birkhäuser Boston, pp. 95-121.

Strange Limit Cycles, Chaos, and Invariant Measure for a Simple Differential-Delay Equation

Uwe an der Heiden, Bremen

INTRODUCTION. Let $a > 0$ and let $f: \mathbb{R} \to \mathbb{R}$ be a real-valued function. The differential-difference equation

$$dx(t)/dt = a \cdot \{ f(x(t-1)) - x(t) \} \tag{1}$$

has been applied by many authors (see [5] for detailed references) to model processes from various areas of biology (e.g. hematopoiesis, enzymatic control, respiration, neural activity) and economics (various types of commodity cycles).

Generally in applications the variable t means time and x some quantity varying with time. The term $-x(t)$ describes the destruction, decay, or efflux of x (hence x has to be nonnegative). The function f describes the production of the quantity x (hence in applications most often $f: \mathbb{R}_+ \to \mathbb{R}_+$ is nonnegative). This production depends on the value of x some time in the past. Therefore we have a system with delayed feedback. The number a can be considered to measure the length of the delay, since by $t' = at$, $y(t') = x(t)$ the equation (1) is transformed into the equation

$$dy(t')/dt' = f(y(t'-a)) - y(t').$$

It is interesting to note that as $a \to \infty$ Eq.(1) in some sense approximates the difference equation

$$x_{n+1} = f(x_n), \quad n=1,2,\ldots \tag{2}$$

which has been widely studied (see e.g. [1] for a survey) because of its rich dynamical behavior. Indeed, as we will clarify here, already for

finite values of a, some properties of solutions to Eq.(1) can be expressed in terms of an equation of type (2).

The behavior of solutions $(x_n)_{n=1,2,\ldots}$ to Eq.(2) is rather
simple if f is a monotone function, but can be very complicated if f has
one or more humps. Mackey and Glass [12] and, independently, Washewska
and Lasota [16] discovered by numerical experiments that similar conditions apply to the solutions $(x(t))_{t>0}$ of Eq.(1). In this paper we discuss
these topics from an analytical point of view, summarizing what has been
proved until now. The proofs of most of the following results are contained in [5] and [4].

2. EARLY RESULTS ON THE EXISTENCE OF PERIODIC SOLUTIONS

One of the most general theorems on the existence of periodic
solutions to Eq.(1) was obtained by Hadeler and Tomiuk [3] . They assumed
the function f and the parameter a to satisfy the following conditions:
(i) f: $\mathbb{R} \to \mathbb{R}$ is continuous and bounded below,
(ii) $f(\xi)\xi < 0$ for all $\xi \neq 0$,
(iii) f is differentiable at O, $f'(O) < 0$,
(iv) $a^2\{(f'(O))^2 - 1\} > \gamma^2$,
where the number γ is determined by $\pi/2 < \gamma < \pi$ and $\gamma\cdot\cot\gamma = -a$ ($\gamma=\pi/2$ if
a=0). Applying an ejective fixed point principle to a cone in the state
space C([-1,0]) they proved that <u>under the conditions</u> (i) <u>to</u> (iv) <u>Eq.(1)</u>
<u>has a non-constant periodic solution with smallest period greater than</u> 2.
Using techniques of phase plane analysis, Kaplan and Yorke [7] obtained
a similar result, additionally assuming some monotonicity condition on f.
Note that (iv) is just the condition for instability of the steady state.
Inspection of the proofs shows that the periodic orbit of these theorems
is simple in the sense that within one smallest period the periodic solution has exacty two extrema (just like the sin-function).

Until now it is not known whether these periodic solutions are
unique and stable (with respect to an appropriate set of initial conditions). However, it turns out that if f is a piecewise constant function or
continuous, but "near" to a step function then many of these and other
problems can be solved.

3. ASYMPTOTICALLY STABLE PERIODIC ORBITS

Let us first consider the simplest nonlinear monotone decreasing step-function given by

$$f_1(\xi) = \begin{cases} c & \text{if } \xi \leq b \\ 0 & \text{if } \xi > b \, , \end{cases} \tag{f_1}$$

where b and c are positive constants.

For $f=f_1$, a solution to Eq.(1) is a continuous function $x: [-1,\infty) \to \mathbb{R}$ satisfying Eq.(1) for all t>0 for which $x(t-1) \neq b$. The fuction $\psi: [-1,0] \to \mathbb{R}$, $\psi(t)=x(t)$ for all $t \in [-1,0]$, is called the initial condition of the solution x. Evidently for each $\psi \in C([-1,0])$ there is a unique solution $x=x_\psi$ to Eq.(1) with $f=f_1$. This solution is piecewise composed of exponential functions, namely for any two times $0<t_0<t$ the following relations hold:

$$x(t) = x(t_0) \exp(-a(t-t_0)) \qquad \text{if } x(s-1)>b \text{ for all } s\in(t_0,t), \tag{3a}$$

$$x(t) = c-(c-x(t_0)) \exp(-a(t-t_0)) \text{ if } x(s-1)\leq b \text{ for all } s\in(t_0,t). \tag{3b}$$

In [5] it is proved that (i) _if $c<b$ then $x\equiv c$ is a globally asymptotically stable solution of Eq.(1) with $f=f_1$_, (ii) _if $c>b$ then Eq.(1) with $f=f_1$ has an asymptotically orbitally stable periodic solution $x=\tilde{x}$. The period of $\tilde{x}$ is larger than 2. The orbit of $\tilde{x}$ attracts all orbits of solutions corresponding to monotone initial conditions. $\tilde{x}$ has exactly one maximum within one smallest period._

These results can also be proved in the case of continuous monotone functions f _near_ to f_1, i.e. functions obeying $f(\xi)=f_1(\xi)$ for all $\xi \notin [b,b+\delta]$, where δ is a positive, but sufficiently small number. For the technique the reader is referred to the paper [6].

4. UNSTABLE PERIODIC SOLUTIONS

The behavior of solutions is quite different, if f is an increasing function (positive feedback). The simplest non-costant step-function of this type is

$$f_2(\xi) = \begin{cases} 0 & \text{if } \xi \leq 1 \\ c & \text{if } \xi > 1 \, . \end{cases} \tag{f_2}$$

If $c<1$ then $x\equiv 0$ is a globally asymptotically stable constant solution of Eq.(1) with $f=f_2$,[5]. If c>1 then there is a second constant solution, namely $x\equiv c$. These steady states are stable. More precisely, _if_ c>1 <u>then</u> <u>both</u> x≡0 <u>and</u> x≡c <u>are</u> <u>locally</u> <u>asymptotically</u> <u>stable</u> <u>constant</u> <u>sol</u>-<u>utions</u> <u>to</u> Eq.(1) <u>with</u> $f=f_2$,[5]. It is not easy to determine the boundary between the domains of attraction of the constant solutions. Obviously solutions corresponding to initial conditions $\psi>1$ converge to c as t → ∞, and if $\psi<1$ then $x_\psi(t)$ → 0 as t → ∞. However, what happens if the initial condition oscillates around the level 1? Consider the following set D of initial conditions: $\psi\in D$ if and only if there is $w=w(\psi)\in[0,1]$ such that $\psi(t)>1$ for all $t\in(-1,-1+w)$ and $\psi(t)<1$ for all $t\in(-1+w,0)$ and moreover $\psi(0)=1$. A continuous map V: D → [0,1] is defined by $V(\psi)=w(\psi)$. It is proved in [6] that <u>for</u> <u>functions</u> $f=f_2$ <u>with</u> c>1 (and also for smooth funct-ions near to f_2 in the sense of Sect.3) <u>there</u> <u>is</u> <u>a</u> <u>unique</u> $w_p\in(0,1)$ <u>such</u> <u>that</u> <u>the</u> <u>following</u> <u>conditions</u> <u>hold</u>:

(i)   <u>if</u> $\psi\in D$, $V(\psi)<w_p$ <u>then</u> $x_\psi(t)$ → 0 <u>as</u> t → ∞,

(ii)   <u>if</u> $\psi\in D$, $V(\psi)>w_p$ <u>then</u> $x_\psi(t)$ → c <u>as</u> t → ∞,

(iii) <u>if</u> $\psi\in D$, $V(\psi)=w_p$ <u>then</u> $\hat{x}=x_\psi$ <u>is</u> <u>periodic</u> <u>for</u> t>0.

Following from (i) and (ii) <u>the</u> <u>periodic</u> <u>solution</u> $\hat{x}$ <u>is</u> <u>unstable</u>. <u>Its</u> <u>period</u> <u>is</u> <u>between</u> 1 <u>and</u> 2.

5. STABLE LIMIT CYCLES OF SPIRAL TYPE

Much more complicated solutions arise if the nonlinearity has mixed feedback characteristics. As a most simple type we consider here the function

$$f_3(\xi) = \begin{cases} 0 & \text{if } 0 \leq \xi < 1 \text{ or } \xi > b \\ c & \text{if } 1 \leq \xi \leq b, \end{cases} \qquad (f_3)$$

where b,c are constants satisfying c > b > 1.

Such a step-function imitates the crucial features of the functions

$$f(\xi) = K \frac{\xi^m}{1+\xi^n} , \quad 1 \leq m < n,$$

originally used in the numerical experiments [2], [12] , leading to the discovery of the complicated dynamics.

Solutions to Eq.(1) with $f=f_3$ again obey the conditions (3a,b). As f_3 is the minimum of f_1 and f_2, it is easy to see that the periodic solutions $\tilde{x}$ and $\hat{x}$ of Sections 3 and 4 are also solutions to Eq.(1) with $f=f_3$ if the thresholds 1 and b are sufficiently apart. More precisely <u>a</u> <u>stable</u> <u>limit</u> <u>cycle</u> $\tilde{x}$ <u>oscillating</u> <u>around</u> <u>the</u> <u>level</u> b <u>exists</u> <u>if</u>

$$b\, e^{-a} > 1, \qquad (4)$$

and $\underline{an}$ $\underline{unstable}$ $\underline{limit}$ $\underline{cycle}$ $\hat{x}$ $\underline{oscillating}$ $\underline{around}$ $\underline{the}$ $\underline{level}$ 1 $\underline{exists}$ $\underline{if}$

$$c > 1 \;\underline{and}\; c-(c-1)e^{-a} \leq b. \tag{5}$$

These cycles are simple in that they have only one minimum within one smallest period. If condition (4) is violated but (5) still holds, then new types of stable limit cycles occur. Roughly speaking, for each positive integer n there are parameters (c,a,b) such that Eq.(1) has a stable periodic solution with n minima per smallest period. Precisely, the following statement has been proved [5]:

$\underline{Let}$ $f=f_3$, $\underline{let}$ z $\underline{be}$ $\underline{the}$ $\underline{positive}$ $\underline{root}$ $\underline{of}$ $\underline{the}$ $\underline{quadratic}$

$$z^2 - (c-(c-1)e^{-a}-c^2)z - e^{-a}c(c-1) = 0, \tag{6}$$

$\underline{and}$ $\underline{let}$ $\underline{the}$ $\underline{inequalities}$

$$c-(c-1)e^{-a} \leq z < c \;\underline{and}\; c(c-1+e^{-a})^{-1} \leq ze^a \tag{7}$$

$\underline{hold}$. $\underline{Then}$ $\underline{there}$ $\underline{is}$ $\underline{a}$ $\underline{sequence}$ $(b_n)_{n=1,2,\ldots}$ $\underline{of}$ $\underline{numbers}$ $b_n{=}b_n(a,c)$ $\underline{satis\text{-}}$
$\underline{fying}$

(i) $c-(c-1)e^{-a}<b_n$, $b_{n+1}<b_n$, $b_\infty = \lim\limits_{n\to\infty} b_n = ze^a$, $b_n \leq \min(e^a,c)$,

(ii) $\underline{for}$ $\underline{each}$ $b\in(b_\infty,b_1]$ $\underline{there}$ $\underline{is}$ $\underline{an}$ $\underline{asymptotically}$ $\underline{orbitally}$ $\underline{stable}$
$\underline{periodic}$ $\underline{solution}$ $x{=}x_b$ $\underline{to}$ $\underline{Eq.}(1)$ $\underline{with}$ $f=f_3$,
(iii) $\underline{if}$ $b\in(b_{n+1},b_n]$ $\underline{then}$ x_b $\underline{has}$ $\underline{exactly}$ $n_o{+}n$ $\underline{minima}$ $\underline{within}$ $\underline{one}$ $\underline{smallest}$
$\underline{period}$, $\underline{where}$ $n_o = n_o(a,c)$ $\underline{does}$ $\underline{not}$ $\underline{depend}$ $\underline{on}$ b,
(iv) $\underline{the}$ $\underline{(smallest)}$ $\underline{period}$ $\underline{of}$ x_b $\underline{converges}$ $\underline{to}$ ∞ $\underline{as}$ $b \to b_\infty$,
(v) $\underline{for}$ $\underline{all}$ $b\in[b_\infty,b_1]$, $\underline{Eq.}(1)$ $\underline{has}$ $\underline{an}$ $\underline{unstable}$ $\underline{periodic}$ $\underline{solution}$ $\hat{x}$ $\underline{not}$
$\underline{depending}$ $\underline{on}$ b,
(vi) $\min\limits_{t} x_b(t) \to \min\limits_{t} \hat{x}(t)$ $\underline{as}$ $b \to b_\infty$, $\max\limits_{t} \hat{x}(t) < b <\max\limits_{t} x_b(t)$.

The proof of this theorem shows that the orbit of x_b with $b\in(b_{n+1},b_n]$ winds $n_o{+}n$ times around the orbit of $\hat{x}$, and therefore may be imagined as a closed spiral with $n_o{+}n$ windings (of different , increasing amplitudes).

6. CHAOS

In 1975 the following theorem (period three implies chaos) of Li and Yorke was published: $\underline{Let}$ I $\underline{be}$ $\underline{some}$ $\underline{interval}$ $\underline{and}$ $G{:}I \to I$ $\underline{a}$ $\underline{contin\text{-}}$
$\underline{uous}$ $\underline{function}$. $\underline{If}$ $\underline{there}$ $\underline{is}$ $r \in I$ $\underline{obeying}$

$$G^3(r) \leq r < G(r) < G^2(r),$$

$\underline{then}$ $\underline{there}$ $\underline{is}$ $\underline{a}$ $\underline{sequence}$ $S_1{=}\{v_1,v_2,\ldots\}$ $\subset I$ $\underline{and}$ $\underline{an}$ $\underline{uncountable}$ $\underline{set}$ $S_2{\subset}I$
$\underline{such}$ $\underline{that}$

(i) $\quad G^k(v_k) = v_k, \; G^i(v_k) \neq v_k$ <u>if</u> $1 \leq i < k$, <u>for all</u> $v_k \in S_1$,

(ii) $\quad \lim\limits_{i \to \infty} \sup \; |G^i(v) - G^i(v')| > 0$

$\qquad \lim\limits_{i \to \infty} \inf \; |G^i(v) - G^i(v')| = 0$ $\qquad$ <u>if</u> $v, v' \in S_2, \; v \neq v'$,

(iii) $\quad \lim\limits_{i \to \infty} \sup \; |G^i(v) - G^i(v')| > 0$ $\qquad$ <u>if</u> $v \in S_1, \; v' \in S_2$. $\hspace{3cm}$ (LY)

Exactly the same type of chaotic motion can be shown to exist for the differential-difference equation (17), if f has certain mixed feedback characteristics. Let us consider a step nonlinearity a little more general than f_3, namely

$$f_4(\xi) = \begin{cases} 0 & \text{if } \xi < 1 \\ c & \text{if } 1 \leq \xi \leq b \\ d & \text{if } b < \xi, \end{cases} \qquad (f_4)$$

where the constants obey $b > 1$, $d \leq c$, and $c > b$. Then the following theorem holds:

<u>Let</u> $f = f_4$. <u>Let the parameters</u> (a,b,c) <u>satisfy</u>

(α) $\qquad b(c-1) < (c-b)e^a$, $\hspace{6cm}$ (9)

(β) $\qquad c^2/(c-1) > b(b-e^{-a})/(b-1)$ $\hspace{4.5cm}$ (10)

(γ) $\qquad \dfrac{c(c-b)(c-1)^2 b}{c(c-b)-b(c-1)\exp(-a)} \leq (2c-1-cb)e^a + (c-1)(c-b)$. $\hspace{1.5cm}$ (11)

<u>Then there is a number</u> $d^* = d^*(a,b,c)$ <u>such that for each</u> $d \leq d^*$ <u>the fol-</u><u>lowing conditions hold</u>:
<u>There is a countably infinite set</u> $T_1 = \{x_k : k=1,2,\ldots\}$ <u>of periodic sol-</u><u>utions and an uncountable set</u> T_2 <u>of aperiodic solutions to Eq.(1) satis-</u><u>fying</u>
(i) <u>to each</u> $x \in T_1 \cup T_2$ <u>there corresponds a sequence</u> $0 < t_1(x) < t_2(x) < \ldots$, $\lim\limits_{i \to \infty} t_i(x) = \infty$, <u>such that</u> $x(t_i(x)) = 1$, $x(t) \neq 1$ <u>for all</u> $t \neq t_i(x)$, $i=1,2,\ldots$,
(ii) <u>there is an interval</u> $I \subset [0,1]$ <u>and a continuous map</u> $G : I \to I$ <u>such</u> <u>that for all</u> $x \in T_1 \cup T_2$

$$t_{2i+2}(x) - t_{2i+1}(x) = G^i(t_{2i}(x) - t_{2i-1}(x)) \;\; \underline{\text{for}} \;\; i=1,2,\ldots, \quad (12)$$

(iii) G <u>obeys</u> (8) <u>and the sets</u> $S_1 = \{t_2(x)-t_1(x) : x \in T_1\}$, <u>and</u> $S_2 = \{t_2(x)-t_1(x) : x \in T_2\}$ <u>obey</u> (LY).

A proof of this theorem together with an exact computation of the function G is contained in [4], with precursors in [6] and [5]. In [6] it is shown how to generalize the result to smooth nonlinearities f near to step functions. Peters[13] and Walther [15] have shown that similar results also hold for the equation $dx(t)/dt = g(x(t-1))$.

Let us note that the initial conditions corresponding to the solutions $x \in T_1 \cup T_2$ are elements of the set D defined above.

7. STATISTICAL BEHAVIOR

Just as a limit cycle or a steady state may be stable or unstable, the chaotic domain in the state space may be stable or unstable. If it is unstable and if moreover its measure in the state space is zero then generally the chaotic orbits will not be observed in any physical realization of the system. Indeed, it can be shown that the chaotic set exhibited by the previous theorem has the structure of a Cantor set, hence has measure zero, and is unstable for broad regions of the parameters (a,b,c,d) satisfying $(\alpha),(\beta),(\gamma)$, and $d \leq d^*$. Therefore it is important to find regions in the parameter space where the chaotic behavior is not exceptional. This problem is considered in the paper [4]. It could be shown under additional assumptions on $f=f_4$ that the map $G:I \to I$ of the previous theorem is invariant with respect to some absolutely continuous measure μ. Moreover G is ergodic, mixing, and exact with respect to μ, and finally all solutions x corresponding to initial conditions $\psi \in D$ with $V(\psi) \in I$ satisfy the relation (12) (where $t_i(x)$ denotes the i-th value of t obeying $x(t)=1$). Before stating the result exactly we briefly report some definitions of statistical notions from the work of Lasota and Yorke [8], [9], [10]. Let I be some bounded interval, A the σ-algebra of Lebesgue-measurable subsets of I, and $\lambda: A \to R_+$ the Lebesgue-measure. A measure $\mu: A \to [0,1]$ is called <u>absolutely</u> <u>continuous</u> with respect to λ if there is a function $f \in L^1(I)$ obeying

$$f \geq 0, \quad \int_I f \, d\lambda = 1, \quad \text{and } \mu(A) = \int_A f \, d\lambda \quad \text{for all } A \in A. \tag{13}$$

A map $G:I \to I$ is called <u>invariant</u> with respect to μ if

$$\mu(A) = \mu(T^{-1}(A)) \quad \text{for all } A \in A. \tag{14}$$

G is called <u>ergodic</u> with respect to μ if
$$A \in A, \ T^{-1}(A) = A \text{ implies } \mu(A) = 0 \text{ or } \mu(I-A) = 0. \tag{15}$$

G is called <u>mixing</u> with respect to μ if T is invariant and if

$$\lim_{n \to \infty} \mu(A \cap T^{-n}(B)) = \mu(A) \cdot \mu(B) \quad \text{for all } A,B \in A. \tag{16}$$

G is called <u>exact</u> if for all $A \in A$ the sets $T^i(A) \in A$, $i=1,2,\ldots$, and if

$$A \in A, \ \mu(A) > 0 \text{ implies } \lim_{n \to \infty} \mu(T^n(A)) = 1. \tag{17}$$

The strongest of these conditions is exactness, since exactness implies

mixing and mixing implies ergodicity (Lasota). Now we state our last
theorem [4], the proof of which is based on results contained in [8]
and [9].

Let $f=f_4$. Assume the parameters a,b,c to satisfy (9),(10), and
(11). Let additionally

$$b^2(c-1)^2 c\left(\frac{d-c}{c-b} + \frac{c}{c-1}\right) < [c^2(b-1) + b(c-1)e^{-a}]^2. \qquad (18)$$

Then for some value of d = d(a,b,c) there is an interval $I \subset [0,1]$ and a
continuous function G: I $\to$ I with the following properties:
(i) for any initial condition $\psi \in D$ obeying $V(\psi) \in I$ the corresponding
solution x to Eq.(1) has an infinite number of values $t_i(x)$, $t_i(x) < t_{i+1}(x)$,
$t_i(x) \to \infty$ as i $\to \infty$, $x(t_i(x))=1$, $x(t) \neq 1$ if $t \neq t_i(x)$, i=1,2,..., and the relation (12) holds,
(ii) there is an absolutely continuous measure μ: A $\to [0,1]$ such that G
is invariant, ergodic, mixing, and exact with respect to μ.

<u>8. REFERENCES</u>

[1] COLLET,P., ECKMANN, J.-P.: Iterated maps on the interval as dynamical systems. Boston: Birkhäuser 1980

[2] GLASS, L., MACKEY, M.C.: Pathological conditions resulting from in-
stabilities in physiological control systems. Annals New York Acad.
Sci. 316, 214-235 (1979)

[3] HADELER, K.P., TOMIUK, J.: Periodic solutions of difference-differ-
ential equations. Arch. Rat. Mech. An. 65, 87-95 (1977)

[4] AN DER HEIDEN, U.: Periodic, aperiodic, and stochastic behavior of
differential-difference equations modeling biological and economical
processes. In: Differential-Differenzengleichungen, Anwendungen und
numerische Probleme (L.Collatz, G.Meinardus, W.Wetterling ,eds.),
Basel - Boston - Stuttgart: Birkhäuser Verlag 1983

[5] AN DER HEIDEN, U., MACKEY, M.C.: The dynamics of production and destruction. J. Math. Biol. 16, 75-101 (1982)

[6] AN DER HEIDEN, U., WALTHER, H.-O.: Existence of chaos in control
systems with delayed feedback. J. Diff. Eqs. 47, 273-295 (1983)

[7] KAPLAN, J.L., YORKE, J.A.: On the nonlinear differential delay equation x(t)=-f(x(t),x(t-1)).J.Diff.Eqs. 23, 293-314 (1977)

[8] LASOTA, A.: Statistical stability of deterministic systems.(in press)

[9] LASOTA, A., MACKEY, M.C.: Probabilistic properties of deterministic
systems. Cambridge University Press, Cambridge-London-New York 1985

[10] LASOTA, A., YORKE, J.A.: Exact dynamical systems and the Frobenius-
Perron operator. Trans. Amer. Math. Soc. 273, 375-384 (1982)

[11] LI, T.Y., YORKE, J.A.: Period three implies chaos. Amer. Math.Month-
ly 82, 985-992 (1975)

[12] MACKEY, M.C., GLASS, L.: Oscillation and chaos in physiological control systems. Science 197, 287-289 (1977)

[13] PETERS, H.: Change of structure and chaos for solutions of $\dot{x}(t) = -f(x(t-1))$. In: Numerical solution of nonlinear equations (E.L. Allgower, K.Glashoff, H.-O.Peitgen, eds.), Springer Lect. Notes in Math. 878, 325-350 (1981)

[14] SHARKOVSKII, A.N.: Coexistence of cycles of a continuous map of a line into itself. Ukr. Mat. Z. 16, 61-71 (1964)

[15] WALTHER, H.-O.: Homoclinic solution and chaos in $\dot{x}(t) = f(x(t-1))$. Nonlinear Analysis 5, 775-788 (1981)

[16] WAZEWSKA-CZYEWSKA, M., LASOTA, A.: Matematyczne problemy dynamiki ukladu krwinek czerwonych. Matematyka Stosowana VI, 23-40 (1976)

The use of Computer Simulation to Evaluate the Testability of a new Fitness Concept

Wilfried Gabriel, Plön

Abstract

Using a model for the evolution of fitness, we show how computer simulation can be used to examine whether a theoretical model can be tested by experimental data. A maximum-likelihood-fit on simulated measurements was able to separate the parameters. Such measurements can be made in a realistic experimental set-up. Therefore, the concept and the predictions of the model can be experimentally tested.

Zusammenfassung

An dem Beispiel eines neuen Fitnesskonzepts, wo der Verlauf der Fitnessfunktion selbst der Evolution unterworfen ist, wird gezeigt, wie Computer-Simulation hilfreich sein kann, um die Testbarkeit von Modellvorstellungen zu untersuchen. In diesem Modellkonzept mit quantitativ genetischen Ansätzen war es ungewiß, ob die wesentlichen Modellparameter überhaupt aus experimentellen Daten bestimmbar sind. Mit einem Maximum-Likelihood-Verfahren gelingt es, diese Modellparameter zu separieren. Simulierte Messungen demonstrieren, daß der dazu notwendige Meßaufwand in realisierbarem Rahmen bleibt. Damit erweisen sich die grundlegenden Modellvorstellungen und Modellvorhersagen als experimentell testbar.

1.   Introduction

The assumptions underlying many mathematical models in biology are such simplifications of reality that the models cannot be tested by empirical data. Nevertheless, such models do clearify ideas and enable precise definitions, as well as promoting a qualitative understanding necessary for the development of new concepts. Realistic models in biology, however, require a set of testable predictions. Crucial experiments or observations should be able to falsify a model or to suggest useful modifications. For that purpose a biological interpretation of all model parameters is required. If these parameters are not directly measurable, a description of a realistic experimental set-up and of a statistical procedure to extract the parameters from data are necessary to achieve creditibility with experimenters.

Here we present an example of a model concept for which it was uncertain, whether the parameters could be estimated from experimental data. In almost all genetic models (e.g. *Lande 1982, Lynch and Gabriel 1983)* the individual fitness function (or equivalently the individual niche width) is fixed and, therefore, independent of evolution. Using concepts from quantitative genetics, we redefine fitness as a tolerance curve which itself changes during the evolutionary process. Adaptation is driven in the direction of maximal fitness. Therefore, the relative fitness contribution of any quantitative trait is calculable by its difference from the possible optimal value and by the shape of the fitness function. However, in temporally and spatially variable environments natural selection changes the breadth of adaptation in relation to expected variations of the optimum. Thus, both the optimum value and the whole shape of a tolerance curve or fitness function are evolving. By incorporating this into a model concept under quite general conditions, the optimal breadth of adaptation can be predicted (most simple in clonal populations) as a function of the variability of the environment *(Lynch and Gabriel 1986a,b)*. Thereby, according to our model, spatial heterogeneity, temporal variation within a generation, and temporal variation between generations act independently. The produced variances represent, in general terms, contributions to the fitness with multiplicative and additive effects.

2. Model Concept

We start by taking the simplest possible assumptions and the minimum number of parameters. Let us assume that the tolerance curve of an individual over an environmental gradient can be represented by two quantitative genetic characters: g_1 describes the optimum and g_2 the variance of the tolerance curve. Each of these characters may be the expression of a large number of genes and is measured on an environmental scale. However, this description is valid only for the average of a clonal population because of the inevitable variation due to development. Therefore, the fitness of an individual is determined rather by its phenotypic values z_1 and z_2, than by the corresponding genotypic values g_1 and g_2. We suppose that in most cases with a proper scale transformation the fitness function can be approximated by a normal distribution:

$$w(z_1,z_2,\phi)=(2\pi z_2)^{-1/2}\exp(-(z_1-\phi)^2/2z_2) . \tag{1}$$

The actual environment (the model is most applicable to physical or chemical gradients like temperature or pH) is measured by ϕ. The variance of phenotypes with identical genotypes growing up under the same conditions is called developmental noise V_E (according to the traditional nomenclature). Quantitative genetics usually puts a normal distribution for the relationship between phenotypes and genotypes

$$p(z_1|g_1) = (2\pi V_{E1})^{-1/2}\exp(-(z_1-g_1)^2/2V_{E1}) \, , \qquad (2)$$

but as z_2 is a variance it always has to be greater than zero. For the distribution of z_2 around g_2 we use a beta distribution of the second kind *(Kendall and Stuart 1977)*

$$p(x) = x^{\alpha-1} (1+x)^{-(\alpha-\beta)}\Gamma(\alpha+\beta)/(\Gamma(\alpha)\Gamma(\beta)) \, .$$

For mathematical reasons it is helpful to set $x = z_2/V_{E1}$. By this substitution V_{E1} acts as a scaling factor. The z_2-distribution, however, remains independent of V_{E1}. From the constraints $E(z_2) = g_2$ and $Var(z_2) = V_{E2}$, it follows that

$$p(z_2|g_2) = (z_2/V_{E1})^{\alpha-1} (1+(z_2/V_{E1}))^{-(\alpha-\beta)}\Gamma(\alpha+\beta)/(V_{E1}\Gamma(\alpha)\Gamma(\beta)) \qquad (3)$$

with

$$\alpha = (g_2(g_2+V_{E1})/V_{E2} +1)g_2/V_{E1}$$

$$\beta = g_2(g_2+V_{E1})/V_{E2} +2 \, .$$

The four variables g_1, g_2, V_{E1}, and V_{E2} are essential for the understanding of the evolutionary process in question. Yet, as shown by the basic equations they have opposing influences on the effective fitness of a population. Therefore, as a first step it is necessary to demonstrate that each of these parameters can be measured under realistic conditions. In order to do this without appropriate experimental data, we simulated experiments with the expected range of variables. The parameter estimation is done by the following rather computer time consuming but effective maximum likelihood procedure.

3. Parameter Identification by the Maximum Likelihood Method

In the following we assume that measures of fitness w_i at environmental states ϕ_i are available for n individuals of a single genotype. The ϕ_i need

not all be unique, but – according to the experience of several simulations –
a minimum of 3-4 environmental settings are recommended in a range where
the fitness changes at least one order of magnitude. Each fitness value is
a function of the corresponding ϕ_i as well as of the unknown parameters
g_1, g_2, V_{E1}, and V_{E2}. We wish to estimate these four unknowns by maxi-
mizing the joint (a posteriori) probabilities of all observations, the product
of all n p_i (or the sum of n $\log(p_i)$), where p_i, the conditional probability
of observing (w_i, ϕ_i), is a function of g_1, g_2, V_{E1}, and V_{E2}.

We will focus on the simple case in which individuals are exposed to constant ϕ throughout their lives as would be approximated in a controlled
laboratory or greenhouse setting. A specific measure of individual i in environment ϕ_i can result from many different combinations of z_1 and z_2, but
equation (1) implies the constraint

$$z_1 = \phi_i \pm (-2z_2\ln(w_i(2\pi z_2)^{1/2}))^{1/2} \qquad . \tag{4}$$

The a posteriori probability of the observation (w_i, ϕ_i) can now be expressed by weighted integration over all possible combinations of z_1 and
z_2. For each z_2 the probability of obtaining the z_1 that results in the
observation (w_i, ϕ_i) can be calculated by substituting equation (4) into (2):

$$H(z_2, w_i, \phi_i) = (2\pi V_{E1})^{-1/2}(\exp(-(x-g_1)^2/2V_{E1})$$
$$+ \quad \exp(-(y-g_1)^2/2V_{E1})) \tag{5}$$

where $x = \phi_i + c$, $y = \phi_i - c$ and
$$c = (-2z_2\ln(w_i(2\pi z_2)^{1/2}))^{1/2} \qquad .$$

The (a-posteriori) probability p_i of the measurement (w_i, ϕ_i) is then given
with (3) and (5)

$$p_i = \int p(z_2|g_2)H(z_2, w_i, \phi_i)dz_2 \qquad .$$

The boundaries of this integration are $z_2 = 0$ and $z_2 = 1/(2\pi w_i^2)$.
Above this limit it is impossible to measure a fitness of the value w_i.
Unfortunately, there is no analytical solution available for this integral.
Another complication arises from the fact that the expectation value of the
geometric mean fitness (f_i) varies slightly with ϕ. For combining measurements of different environmental settings, the contributions of distinct ϕ
values to the joint probability have to be weighted by the expectation
values f_i in order to avoid systematical errors. Given the above equations

the maximum likelihood estimates of the four parameters g_1, g_2, V_{E1}, and V_{E2} are obtainable using well-known Monte-Carlo-techniques for f_i, numerical integration for p_i and optimization procedures to maximize the joint probability. At the end of the maximization, estimates of the second derivatives at the maximum may also be computed to obtain (from the inverse of the Hessian matrix) estimates of the (sampling) variances and covariances of the four parameters.

This same analysis could be applied in situations in which ϕ is not constant within the lifetime of measured individuals but has for all ϕ settings the same variance V_ϕ. In this case, using equations derived in *Lynch and Gabriel (1986b)*, the equation (4) becomes

$$z_1 = \phi_i \pm (-2z_2 \ln(w_i(2\pi z_2)^{1/2})-V_\phi)^{1/2} . \tag{6}$$

With appropriate sample size, use of (6) in the definition of p_i would enable the investigator to derive an estimate of V_ϕ as well as g_1, g_2, V_{E1}, and V_{E2}.

4.   Results

The dotted line in the figure represents the expected fitness distribution at five environmental values (ϕ = 0, 2.23, 5.53, 8.3, and 11.07) for a given parameter set (g_1=0, g_2=10, V_{E1}=1, and V_{E2}=3). The scale of the environmental variable ϕ is chosen so that the fitness function of an average animal has its maximum at ϕ=0. Besides the different numerical range the form of the distribution of fitness values changes with ϕ and shows distinct asymmetries. It can be shown that these distributions and especially the asymmetries depend characteristically on the parameters g_1, g_2, V_{E1}, and V_{E2}. Therefore, it is necessary to have the measurements of single individuals; the usually published mean values and standard deviations of fitness are not sufficient.

The figure also contains simulated measurements. These data are presented as histograms in order to faciliate graphical presentation and comparisons with expected values. For each of the five ϕ-values we have simulated 40 measurements. The procedure described above was then applied to the joint distribution of all 200 data points. The results on the presented set of simulated data are g_1=0.51, g_2=8.85, V_{E1}=1.14, and V_{E2}=3.30. The

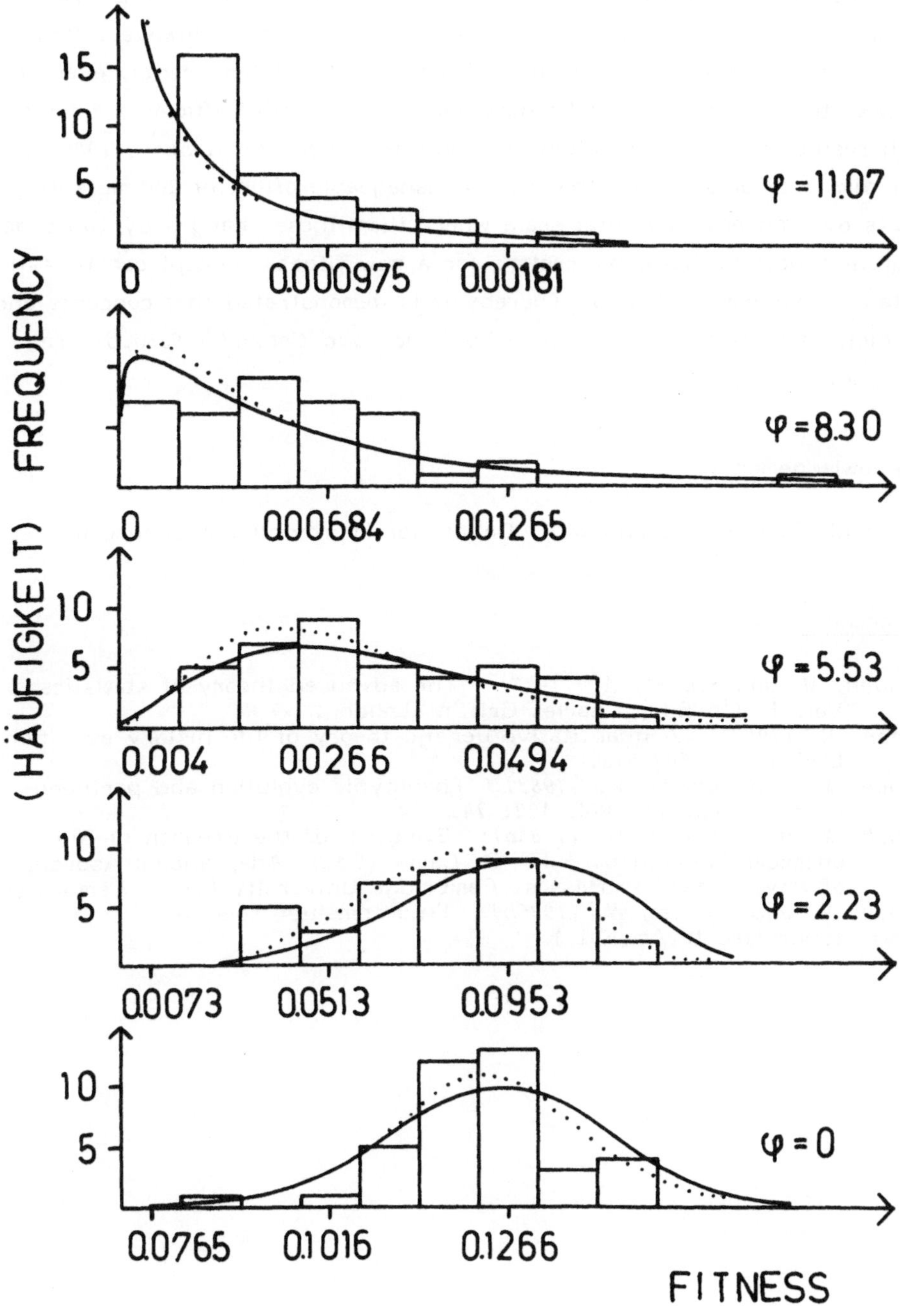

FREQUENCY
(HÄUFIGKEIT)
FITNESS
φ =11.07
φ =8.30
φ =5.53
φ =2.23
φ = 0
15
10
5
0
0.000975
0.00181
0
0.00684
0.01265
10
5
0.004
0.0266
0.0494
10
5
0.0073
0.0513
0.0953
10
5
0.0765
0.1016
0.1266

expected fitness distribution according to this maximum likelihood result
are given by the unbroken line in the figure. Our shown results are
atypical in that they appear to fit the data only poorly. However, this
may arise from the use of too few statistics of simulated events, even with
respect to the dotted line (the goodness of fit is still sufficient; a table
with results of other simulations is given in *Lynch and Gabriel, 1986b*).
Our aim is to demonstrate that with a manageable effort of 200 measurements over an environmental scale where the fitness changes by two orders
of magnitude, the basic parameters for a new fitness concept can be estimated by experimental data. Thereby it is demonstrated that concepts and
predictions of this model as given by *Lynch and Gabriel (1986a,b)* are
testable.

Acknowledgements

We thank G. Carvalho and W.R. DeMott for improving the manuscript.

References

Kendall, M. and Stuart, A. (1977): The advanced theory of statistics,
    Vol. 1, 4th Ed., Charles Griffin, London.
Lande, R. (1982): A quantitative genetic theory of life history evolution.
    Ecology 63: 607–615.
Lynch, M. and Gabriel, W. (1983): Phenotypic evolution and partheno-
    genesis. Am.Nat. Vol. 122: 745–764.
Lynch, M. and Gabriel, W. (1986a): Evolution of the Breadth of Bio-
    chemical Adaptation. – In: P. Calow (Ed.): Adaptational Aspects
    of Physiological Processes. Cambridge University Press, Cambridge.
Lynch, M. and Gabriel, W. (1986b): Environmental Tolerance.
    (submitted to Am.Nat.)

Texture Analysis using Random Field Models exemplified on Ultrasonic Images of the Liver

Ulrich Ranft, Hannover

Zusammenfassung: Anhand eines Beispiels aus der Medizin soll der praktische Einsatz von Modellen zur Bildanalyse demonstriert werden. Hierzu wird ein zweidimensionales stochastisches Modell, das sogenannte autoregressive periodische Modell, zur Texturanalyse vorgestellt. Die Anwendung bildererzeugender Modelle erlaubt die Methode der "Analyse durch Synthese". Indem das Modell einem gegebenen Texturmuster angepaßt wird, stehen mit den geschätzten Modellparametern geeignete Texturmerkmale zur Verfügung. Seine erfolgreiche Anwendung zur Quantifizierung der Echomuster von Ultraschallbildern der Leber wird gezeigt.

Summary: With the aid of a medical example the practical use of models for image analysis will be demonstrated. For that purpose a random field model, the so-called autoregressive periodic random field model, for texture analysis is briefly described. The use of a generative model offers the analysis by synthesis approach. By fitting the model to a given texture pattern the estimated model parameters are suitable textural features. Its successful application to the quantification of the echo pattern of ultrasonic images of the liver can be shown.

Introduction

Image models are now becoming more and more an important tool in image analysis like simulation models in the analysis of biological systems and processes. The central interest of image modeling can be reduced to giving a description of observed images in terms of algebraic and probalistic processes that generate the images. This viewpoint of image modeling comprises a broad spectrum of image models, e.g. syntactic, stochastic, or boolean models [1]. The image model may or may not reflect the real image formation process. The practical use of image models will be demonstrated on the quantification of echo patterns of ultrasonic images of the liver by using random field models (RFM). Besides the important structural information such as size and shape of tumoreous regions, an ultrasonic image of the liver carries also textural information which is useable for characterization of the liver tissue, e.g. as normal or pathological. By fitting a RFM to a region of interest (ROI) of homogeneous texture pattern the estimated model parameters are suitable textural features.

In the following section for a specific type of RFMs, the autoregressive model on a regular finite lattice, the model equations and procedures for image synthesis and parameter estimation are briefly described. In the third section, the concept of texture analysis by means of RFMs for tissue characterization is explained, and some results are given. Concluding remarks stress the demonstrated advantages of image modeling for image analysis.

Autoregressive RFMs on Regular Finite Lattices

First practical attempts of texture analysis using stochastic models have been simply the application of time series models to images [2]. BESAG [3] developed an extended theory for spatial interaction models of lattice systems, also called as random field models. Especially for image analysis, the RFMs defined over regular lattices are of interest. From the viewpoint of synthesis and analysis KASHYAP et al. [4] demonstrated the advantage of the so-called periodic RFMs. The presented application of RFMs to texture analysis confines itself to the subclass of autoregressive periodic models, which does not mean any principal restriction.

Let the mxm image

$$I_m = \{y(s); s\epsilon\Omega\}, \text{ where } \Omega = \{(i,j); 0{\leq}i,j{\leq}m-1\} \text{ and}$$

$$y(s) \text{ is the pixel intensity,}$$

so the periodic assumption can be expressed by

$$y(s+q) = y(s)\epsilon I_m, \text{ where } s\epsilon\Omega \text{ and } q = (k(m+1),l(m+1)) \text{ and } k,l=0,\pm1,\pm2,\ldots$$

A central part in the theory of spacial interaction models plays the neighborhood set N of a pixel with the meaning that the pixel intensity $y(s)$ is closely related to a weighted sum of the intensities of a finite number of neighboring pixels:

$$N = \{r_i = (r_{i1},r_{12}){\neq}(0,0); i=1,\ldots,n \text{ and } r_{ij} \text{ integers}\}$$

A autoregressive periodic RFM can now be defined by

$$(1) \quad y(s)-\alpha = \sum_{r\epsilon N} \vartheta(r) [y(s+r)-\alpha] + \sqrt{\rho}u(s), \text{ where } s\epsilon\Omega \text{ and}$$

$$\{u(s)\} \text{ is a sequence of iid random variables with } E[u(s)]=0,$$

$$Var[u(s)]=1, \text{ and } Cov[u(s_1)u(s_2)]=0 \text{ for } s_1{\neq}s_2, \text{ and } \alpha=E[y(s)].$$

To ensure stationarity, the coefficients $\vartheta(r)$ must obey

$$(2) \quad \sum_{r\epsilon N} \vartheta(r) \exp(2\pi i/m \; r{\cdot}s) \neq 1 \text{ for all } s\epsilon\Omega.$$

The n+2 model parameters $\{\alpha, \vartheta(r), \rho\}$ completely define a specific texture pattern. Given a parameter set and a set of random variable realizations $\{u(s)\}$, equation (1) specifies in terms of a convolution the instruction for image synthesis. On the other hand, if an observed image is given, the estimated parameters represent, within the means of the model, the texture pattern of the observed image. For both the image synthesis and the parameter estimation computationally simple procedures can be given.

Because of the periodic assumption the convolution in equation (1) becomes after two dimensional discrete Fourier transformation a simple multiplication:

(3) $U(w) = 1/\sqrt{\rho} \; \Theta(w) \; Y(w)$, where $w \in \{(u,v); \; 0 \leq u, v \leq m-1\}$,

   U and Y are the discrete Fourier transforms of u and y-α, respectively, and

   $$\Theta(w) = 1 - \sum_{r \in N} \vartheta(r) \; \exp(2\pi i/m \; r \cdot w) \neq 0.$$

Therefore, an image generation process runs in five steps, i.e. generation of mxm random numbers with zero mean and unit variance, two dimensional discrete Fourier transformation of the random image, pixel by pixel multiplication of the transformed image according to (3), inverse discrete Fourier transformation, and, finally, a correction for the image mean value α. If one restricts the neighborhood N to be unilateral, i.e. if $(i,j) \in N$ then $(-i,-j) \notin N$, and the distribution of the random variables $u(s)$ to be Gaussian, then the least-square estimates turn out to be close approximations of the maximum likelihood estimates which are the optimal ones:

$$\hat{\underline{\vartheta}} = [\sum_s \underline{z}_s \underline{z}_s^T]^{-1} \sum_s y'(s)\underline{z}_s \;, \quad \hat{\rho} = \frac{1}{m^2} \sum_s (y'(s) - \hat{\underline{\vartheta}}^T \underline{z}_s)^2,$$

   where $\hat{\underline{\vartheta}} = [(r_1),\dots,(r_n)]^T$, $\underline{z}_s = [y'(s+r_1),\dots,y'(s+r_n)]^T$,

   $y'(s) = y(s) - \hat{\alpha}$, and $\hat{\alpha} = \frac{1}{m^2} \sum_s y(s)$

Since only an one-pass through the image and simple computational operations are necessary, a fast computer implementation of the estimation algorithm is feasible.

Texture Analysis of Ultrasonic Images of the Liver

Efforts of a quantitative evaluation of the echo signal for tissue characterization are as old as ultrasound in medicine itself. In principle, two approaches or a mixture of both are possible: On the one hand, the analysis of the physical phenomena of the interaction between tissue and ultrasound, and, on the other hand, the statistical pattern analysis of the ultrasonic echo signals. Whereas the physical approach is still far from a successful use in the clinical routine, results of the statistical approach for the liver tissue characterization, for instant, are encouraging [5]. In the conventional statistical approach, so far used in the literature, a large number of statistical parameters describing textural properties, up to several hundred, make up a pool from which the relevant parameter sets for tissue characterization are selected by means of heuristical criteria and discriminant analysis. RFMs for texture analysis offer the advantages of comparatively small parameter sets and of the image regeneration to value visually the adequacy of the selected model for the texture analysis of a specific image or class of images. If the model type of an autoregressive RFM has been chosen as described in the foregoing section, the selection of the appropriate model simply reduces to the choice of the neighborhood N. In a learning phase the visual criterion takes an important place for neighborhood selection besides a model based criterion [6] and the discriminant analysis.

To apply the autoregressive model for characterizing the echo pattern in ultrasonic images of the liver, the ROI of an image has to satisfy the two conditions of isotropy of spacial resolution and of a sound, free from artefacts representation of the liver parenchyma, respectively. The real time parallel scan ultrasound echo system used in this study fulfils these conditions by choosing interactively a small ROI which represents a parenchyma cut of a few square centimeter [7]. For a learning sample of 60 ROIs, including normal and pathological cases, a neighborhood set of about 15 pixels provides acceptable model fits (see figure). A linear discriminant analysis yields a reclassification rate of 97 % and a leave-one-out classification rate of 85 % for correct classifying in the normal and pathological groups, respectively. In a comparison study with the same data and a set of conventional texture parameters, the so-called Haralick-parameters [8], comparable classification results are obtained.

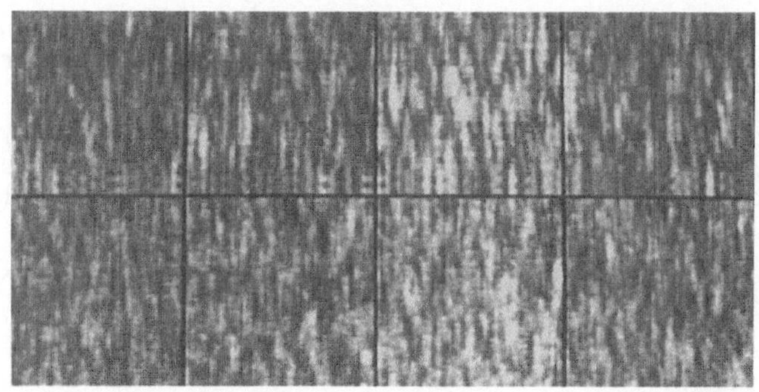

Figure: 4 ROIs of ultrasonic B-scan images of the liver (horizontal beam
direction), 1st row: original echo pattern, 2nd row: regenerated
echo pattern by means of an autoregressive RFM

Conclusions

It could be shown that the estimated parameters of an autoregressive RFM
are just as effective as conventional texture features for tissue characterization of ultrasonic liver images. But, several important characteristics of the model approach are conspicuous. The number of texture features
are kept to a comparatively low amount, but the feature set is still complete within the means of the model. The use of a generative model offers
the analysis by synthesis approach which includes the excellent test by
visual inspection for determining whether the texture has been appropri-
ately represented by the given model. As significant with respect to a
routine application, computationally very efficient procedures for model
parameter estimation and image regeneration are available.

References

[1] Rosenfeld, A.; ed.: "Image Modeling". Academic Press, New York, 1981

[2] McCormick, R.; Jayaramamurthy, S.N.: Time series model for texture
synthesis. Int. J. Comp. Sci. 3, 329-343 (1974)

[3] Besag, J.: Spacial interaction and the statistical analysis of lattice
systems. J. R. Statist. Soc. B 36, 192-236 (1974)

[4] Kashyap, R.L.; Chellappa, R.; Khotanzad, A.: Texture classification
using feature derived from random field models. Pattern Recognition
Letters 1, 43-50 (1982)

[5] Raeth, U.; Schlaps, D.; Limberg, B.; et al.: Diagnostic accuracy of
computerized B-scan texture analysis and conventional ultrasonography
in diffuse parenchymal and malignant liver disease. J. Clin. Ultrasound
13, 87-99 (1985)

[6] Kashyap, R.L.; Chellappa, R.: Estimation and choise of neighbors in spacial-interaction models of images. IEEE Trans. Inf. Theory IT-29, 60-72 (1983)

[7] Ranft, U.: Zweidimensionale stochastische Modelle zur Klassifikation von Echomustern in Ultraschall-parallel-Scan-Bildern der Leber. "Mustererkennung 1985", H. Niemann (ed.), Springer-Verlag, Berlin, 1985, pp. 65-68

[8] Haralick, R.M.: Statistical and structural approaches to texture. Proc. IEEE 67, 786-804 (1979)

Smooth Descriptive Modelling of Multifactorial Systems Responses

Gerd H. Klein, Osnabrück

Summary. Explanatory modelling of complex systems often leads to subsystems responses which have to be modelled descriptively. Wherever nothing is known about such a response except that it may be expected to depend smoothly on its relevant factors, it can conveniently be modelled by a pseudo-cubic spline function. Necessary data input is a set of empirical response values at factor value combinations, scattered in the domain where the spline is to be used for response prediction. The technique is supported by a portable program toolset. Current applications to seasonal modelling of carbon flux in terrestrial biota are sketched.

1. Introduction

Computer simulation gives insights into complex systems behaviour which are particularly useful in forecasting. Reliability depends of course on quality of the underlying mathematical model. Usual modelling approach is explanatory [13]. The interesting response of a system is explained by the interaction of subsystems, and their respectively relevant responses to respectively relevant factors. Sooner or later, iteration of this reduction will stop at subsystems which could be termed "opaque" or "irreducible", with respect to the interesting responses. Reasons could be missing knowledge or missing budget.

Such subsystems responses are usually modelled descriptively, starting from expert knowledge, or assumption, of a complete set of relevant factors which affect the interesting response. Empirical database is noisy values of this response at a set of "spots", that is, value configurations of the factors. A function F is then sought which approximates the true response function Φ, within the region of the spot cloud.

Matrix notation is convenient here. Value configuration of d factors, in the respective units, is denoted $x=(x_1,\ldots,x_d)'$. (The dash denotes transposition, i.e., interchange of rows and columns.) Standard Euclidic vector norm is $\|x\|^2=x_1^2+\ldots+x_d^2$. Noisy values of Φ at the data spots $z^{(i)}$ are denoted $y^{(i)}$

for i=1,...,m. Introducing a proper ststistical weight matrix W, as specified below, and vectors $y=(y^{(1)},...,y^{(m)})'$ and $u[F]=(F(z^{(1)}),...,F(z^{(m)}))'$, F is demanded to minimize, within a properly chosen set of admissible functions, the unfidelity to the data,

$$R_0[F]=(y-u[F])'W(y-u[F])$$

A "mechanistic" [9] parametrization of F, sufficiently justified by prior knowledge about the response mechanism, is not available for opaque systems. Depending on the field of application, there are apparently preferences for certain ad-hoc parametrizations, as exemplified in section 4. By chance, these may work quite well but permanent danger is that a bad parametrization of F may devaluate good empirical data.

In addition, even with a good mechanistic parametrization, necessity of non-linear parameter estimation is often a practical obstacle [1]. We are looking for techniques which, from the data alone, generate handy expressions for F. It will only be assumed that Φ is sufficiently smooth, in some sense; and it will come out that flexible and formally simple parametrizations exist which can be treated by <u>linear</u> parameter estimation where a unique solution is guaranteed under weak conditions.

For conciseness, standard multiindex notation [12] is used: $\mu=(\mu_1,...,\mu_d)$ with non-negative integer components is a d-index with norm $|\mu|=\mu_1+...+\mu_d$. Abbreviations $x^\mu=x_1^{\mu_1}\cdot...\cdot x_d^{\mu_d}$, $\mu!=\mu_1!\cdot...\cdot\mu_d!$, and $D^\mu=\partial^{|\mu|}/\partial x_1^{\mu_1}..\partial x_d^{\mu_d}$ are then very convenient. Script symbol $\mathcal{F}$ will denote Fourier transformation and $\mathcal{E}$ statistical ensemble averaging.

2. Non-splines

A tempting parametrization of F is linear combination of suitable functions,

$$F(x)= \sum_{i=1}^{n} a_i A_i(x)$$

with $n \geq m$. With $a=(a_1,...,a_n)'$ and an mxn-matrix A with elements $A_{ik}=A_k(z^{(i)})$ one has $u[F]=Aa$ and thus $R_0[F]$ is minimized by [1]

$$a=(A'WA)^{-1}A'Wy$$

provided A'WA is regular (and not too severely ill-conditioned). The crucial problem is here the selection of proper basis functions A_i. Parametrization

$$F(x) = \Sigma b_\mu B_\mu(x)$$
$$B_\mu(x) = B_{\mu_1}(x_1) \cdot \ldots \cdot B_{\mu_d}(x_d)$$

where multiindex μ runs in the d-grid $\{1,\ldots,n_1\} \times \ldots \times \{1,\ldots,n_d\}$, is the well-known tensor-product expression which is obviously impracticable for large d, due to the vast number of parameters b_μ. For $B_\mu(x)=x^\mu$, a promising "self-organizing" or "cybernetic" method has been described recently [3] which selects reasonably few significantly contributing d-monomials. Performance quality remains to be seen.

Very flexible and explicitly adapted to expected smoothness is local fitting of the p-th order Taylor expansion of Φ around the interesting point x [12]:

$$F(x+t) = \sum_{k=0}^{p} \sum_{|\mu|=k} x^\mu D^\mu \Phi(x)/\mu!$$

Weight matrix in $R_0[F]$ is chosen here as $W=W(x)=\text{diag}(w_1(x),\ldots,w_m(x))$, with distance dependent weights $w_i(x)=Q(\|x-z^{(i)}\|)$. Values of the partial derivatives $D^\mu \Phi(x)$ are chosen as to minimize $R_0(x)$, by standard linear estimation [1]. The idea is to stabilize the estimator $F(x)$ of $\Phi(x)$ by simultanous optimization of the higher derivatives. Special case $p=d=1$ is propagated in [7]. For $p=0$, the minimizer of $R_0[F]$ is the well-known (generalized) Shepard expression [4],

$$F(x) = \sum_{i=1}^{m} w_i(x)y^{(i)} / \sum_{i=1}^{m} w_i(x)$$

When bulk contribution to $R_0[F]$ stems from a d-sphere $S_r(x)$ of radius r around x, the approach is only consistent if, within $S_r(x)$, Φ is closely approximated by the above Taylor expansion. This is a serious practical problem.

For example, when $Q(s)=\exp(-s^2/h^2)$, radius r is related to the parameter h which controls the extent of smoothing. As yet, there is not statistical technique for estimation of the proper value of h. This is particularly problematic when spot design is clustered and p is small.

An additional practical shortcoming of local expansion techniques is their need for the complete data spot set for any interesting x. For large m, the model function is thus practically incommunicable. A parametrization which combines both flexibility and compactness is desirable. Non-tensor-product spline functions fulfill these requirements. A particularly useful family of such splines has been introduced recently [3] in a very formal function space setting, and already successfully applied practically [14].

3. Splines

Expecting that Φ is smooth, a proper low pass filter $V[F]$ is added to $R_0[F]$ in order to suppress unwanted undulations of F:

$$R_\lambda[F] = R_0[F] + \lambda V[F]$$

with a non-negative Lagrangian multiplier λ. When the noise $\delta = y - u[\Phi]$ is un-biased, $\mathcal{E}(\delta) = 0$, a natural choice for the weight matrix in $R_0[F]$ is $W^{-1} = \mathcal{E}(\delta\delta')$. A formally very simple spline results from choosing

$$V[F] = \sum_{|\mu|=2} \int_{\mathbb{R}^d} \| x \| \, |\mathcal{F} D^\mu F(x)|^2 dx$$

Smoothing action of $V[F]$ may be readily verified using $\mathcal{F} D^\mu F(x) = (-\sqrt{-1})^{|\mu|} x^\mu \mathcal{F} F(x)$: large values of $|\mathcal{F} F(x)|^2$ are less penalized when x is near the origin, that is, in the low frequency domain. Under weak conditions on data spot design, $R_\lambda[F]$ is minimized by the pseudo-cubic spline [2,14],

$$F(x) = \sum_{i=1}^{n} c_i \|x - x^{(i)}\|^3 + b_{d+1} + \sum_{k=1}^{d} b_k x_k$$

Proper choice of the $n \leq m$ spline "knots" $x^{(i)}$ is described below. Parameter vectors $c = (c_1, \ldots, c_n)$ and $b = (b_1, \ldots, b_{d+1})$ depend on λ which controls the extent of smoothing. Computation of c and b is described elsewhere [6], to-gether with statistical techniques for estimating, from the data alone, the correct value of λ.

The spline knots must not be confined to a subspace of Euclidean d-space, i.e., they must at least contain the d+1 edges of a non-degenerate d-simplex (a triangle for d=2, a tetrahedron for d=3). The same applies to the data spots. It will be favourable when the knot domain somehow matches the spot domain. Neither the spots not the knots need to be scattered homogenously within a convex region of d-space.

As few knots as possible should be used for the spline which is uniquely de-fined by (n+1)(d+1) real numbers. Simplest approach to automatic knot place-ment, for a given data spot design, is global specification of a minimum tolerable knot-to-knot distance and a maximum tolerable knot-to-spot-set distance. A stochastically generated trial point is accepted as an additional knot when it is tolerable in this sense. Slight adaption of nearest knot-to-knot approach to local curvature of Φ may be reasonable. Necessary local estimates of $\partial^2\Phi/\partial x_i \partial x_k$ may be computed along the lines of section 2.

Necessary knot number depends of course on the degree of smoothness of Φ; it is also roughly proportional to the volume of the data spot cloud. There are many favourable practical cases where Φ is extremely smooth and where spot cloud volume is small as compared to the volume of its convex hull. Relatively few knots are then sufficient even when d is large. This makes pseudo-cubic splines so useful in smooth descriptive modelling.

4. Applications

Development of the above-described techniques was triggered by experiences in a running plant-ecological project. Carbon flux between terrestrial biota and atmosphere is to be simulated, timespace resolution being typically counties and months, or less. The project contributes to global carbon cycle modelling and research on possible global climatic consequences of increasing atmospheric carbon dioxide content.

Necessary degree of explanatory reduction is still under debate: descriptive modelling could start on the ecological, the ecophysiological, or physiological scale. Anyway, respectively interesting subsystems responses will depend on several respectively relevant factors. This makes versatile multifactorial tools necessary which support modellers in playing with different model assumptions and studying their consequences.

As an example, several factors contribute to photosynthetic production of grassland: light intensity, air and soil temperature, atmospheric carbon dioxide concentration, water availability, and seasonal factors. Favourite ad-hoc parametrizations for this response are apparently $F(x)=F_1(x_1)\cdot\ldots\cdot F_d(x_d)$ and $F(x)=\min\{F_1(x_1),\ldots,F_d(x_d)\}$ [5] which seem to be motivated by Liebig's famous law of the "limiting factor", as originally meant for agraric yield responses. Spline modelling is felt to be a more natural approach here. Reconstruction of test functions, starting from empirical 6-dimensional data spot designs provided by our experimental group, gave very satisfying results.

Seasonal factors deserve special attention here. The length of a certain phenophase, for example dormancy, flowering, senescence, is a kind of medium-term memory of vegetation units on the history of environmental factors. In analogy to prior approaches [10], phenophase durations will be modelled descriptively. An empirical approach [8] to phenological forecasting is being accordingly reformulated and generalized to include factors other than temperature, using again pseudo-cubic splines for these submodels.

References

[1] Y. Bard, Nonlinear Parameter Estimation (Academic, New York, 1974) 58-61, 120-123.

[2] J. Duchon, Splines minimizing rotation-invariant semi-norms in Sobolev spaces, in: W. Schempp and K. Zeller, eds., Constructive Theory of Functions of Several Variables (Springer, Berlin, 1977) 85-100.

[3] S.J. Farlow, The GMDH algorithm, in: S.J. Farlow, ed., Self-Organizing Methods in Modeling (Dekker, New York, 1984) 1-24.

[4] W.J. Gordon and J.A. Wixom, Shepard's method of metric interpolation to bivariate and multivariate data, Math. Comp. 32 (1978) 253-264.

[5] A.W. King and D.L. DeAngelis, Information for Seasonal Models of Carbon Fluxes in Terrestrial Biomes, Publication no. 2485, Environmental Sciences Division, Oak Ridge National Laboratory (1985) 28-32.

[6] G. Klein, Fitting simple non-tensor-product splines to scattered noisy data on Euclidean d-space, submitted for publication (1986).

[7] J. Peil and S. Schmerling, Die lokal angepaßte Regression und ihre Anwendungsmöglichkeiten, Gegenbaurs morph. Jahrb. 126 (1980) 221-227.

[8] A.S. Podolsky, New Phenology: Elements of Mathematical Forecasting in Ecology (Wiley, New York, 1984).

[9] M.S. Rao and S.S. Iyengar, Applications of statistical techniques in modeling of complex systems, in: S.S. Iyengar, ed., Computer Modeling of Complex Systems (CRC, Boca Raton, 1984) 29-53.

[10] R.H. Sauer, A simulation model for grassland primary producer phenology and biomass dynamics, in: G.S. Innis, ed., Grassland Simulation Model, Ecological Studies Vol. 26 (Springer, New York, 1972) 55-87.

[11] L.L. Schumaker, Fitting surfaces to scattered data, in: G.G. Lorentz, C.K. Chui, and L.L. Schumaker, eds., Approximation Theory II (Academic, New York, 1976) 203-268.

[12] L.L. Schumaker, Spline Functions: Basic Theory (Wiley, New York, 1981) 501-523.

[13] J.A. Spriet and G.C. Vansteenkiste, Computer-aided Modelling and Simulation (Academic, London, 1982) 12-55.

[14] G. Wahba, Surface fitting with scattered noisy data on Euclidean d-space and on the sphere, Rocky Mountain J. Math. 14 (1984) 281-299.

Zur Beschreibung offener thermodynamischer Prozesse durch Bindungsdiagramme

Armin Schöne, Bremen

Zusammenfassung: Nach einer Einführung in die bisherigen Ansätze zur Beschreibung thermodynamisch-verfahrenstechnischer (offener thermodynamischer) Prozesse mittels Bindungsdiagrammen wird eine allgemeine Methode zur Umformung von Zustandsdifferentialgleichungen in Bindungsdiagramme angegeben. Als Beispiel wird das Bindungsdiagramm für einen in einem ideal durchmischten Rührkesselreaktor ablaufenden nichtlinearen dynamischen chemischen Prozeß entwickelt.

Summary: The known attempts of describing chemical engineering (open thermodynamic) processes by bond graphs are introduced shortly. Thereafter a general method for transforming state space differential equations into bond graphs is given. The bond graph for a nonlinear dynamic chemical process in a stirred batch reactor with perfect mixing is developed as an example.

1. Einführung

Bindungsdiagramme (Leistungs-Bindungsdiagramme, bond graphs) gehen auf
Paynter /1/, Karnopp und Rosenberg /2/, /3/ zurück. Sie dienen zur Beschreibung dynamischer Prozesse und der Systeme, in denen diese Prozesse ablaufen, auf der Grundlage bestimmter konzentrierter Elemente und sonstiger
Vereinbarungen. Die konzentrierten Elemente sind durch "Bindungen" verknüpft,
denen jeweils zwei Variable, Spannungsvariable und Flußvariable genannt,
zugeordnet sind. Den beiden Variablen werden entgegengesetzte kausale
Wirkungsrichtungen zugewiesen. Außerdem ist für jede Bindung die positive Richtung des Produktes dieser beiden Variablen zu spezifizieren. Nach dem ursprünglichen Konzept der Bindungsdiagramme sollte dieses Produkt eigentlich einen

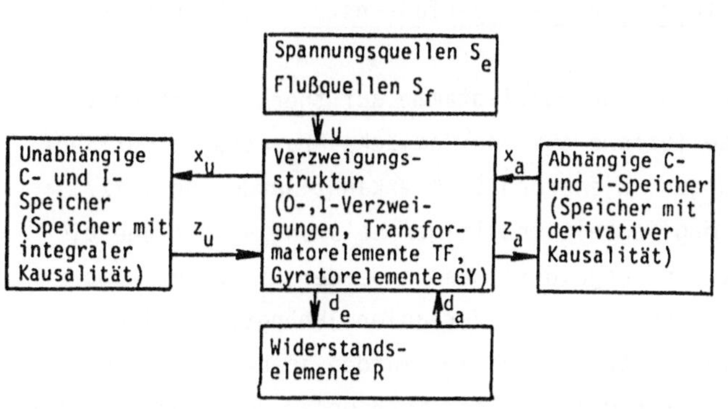

Bild 1. Allgemeine Struktur von Leistungs-Bindungsdiagrammen

Energiestrom (eine Leistung) darstellen. Allgemein kann man den Aufbau eines
derartigen Bindungsdiagrammes gemäß Bild 1 gliedern (u.a. /3/, nähere Er-
läuterung in /4/).

Bei der Entwicklung eines Bindungsdiagramms für einen vorgegebenen Prozeß
kann man im allgemeinen unterschiedliche "Energieregionen" unterscheiden,
z.B. mechanische, elektrische, hydraulische Energieregionen. Nach den ursprünglichen Vorstellungen, die Bindungsdiagrammen zugrunde liegen, sollte
innerhalb einer Energieregion an allen Bindungen die gleiche Art der Flußvariablen und die gleiche Art der Spannungsvariablen eingeführt werden,
also z.B. in einem elektrischen Prozeß (System) der elektrische Strom als
Flußvariable und die elektrische Spannung als Spannungsvariable. Eine solche
a priori-Festlegung engt freilich die Möglichkeit, Prozesse durch Bindungsdiagramme darzustellen, ein.

Anwendungen von Bindungsdiagrammen zur Beschreibung mechanischer, elektri-
scher und hydraulischer Prozesse wurden - unter gewissen einschränkenden
Voraussetzungen - bereits von Karnopp und Rosenberg weit entwickelt. Die
Vorzüge einer Prozeßbeschreibung mittels Bindungsdiagrammen liegen darin,
daß man von vornherein - im Rahmen der vorgegebenen Definitionen der Bindungsdiagramme - zu einer Strukturierung des beschriebenen Prozesses einschließlich seines dynamischen Verhaltens gelangt, die gegebenenfalls auf
einer Beschreibung der Energieströme im betrachteten Prozeß (System) auf-
baut. Obendrein erleichtert diese Prozeßbeschreibung die numerische Bestimmung des dynamischen Verhaltens sehr, da sie sich z.B. direkt als Ein-
gabe in bestimmte Simulationsprogrammsysteme verwenden läßt, u.a. /5/ - /7/.

2. Bindungsdiagramme für thermodynamisch-verfahrenstechnische Prozesse

Die Anwendung der Methode der Bindungsdiagramme auf thermodynamisch-verfahrenstechnische Prozesse (offene thermodynamische Prozesse) unterliegt
weitergehenden Beschränkungen, wie schon früh erkannt wurde, vgl. /2/,
S. 14. Dort wird auf die Möglichkeit ungültiger Leistungs- oder Energie-
interpretationen verwiesen und angeführt, daß z.B. der mit einer Dampf-
leitung verbundene Energiestrom von drei Variablen abhängt, nämlich von
Druck, Volumenstrom und Temperatur. Insgesamt wurden in der Literatur
unterschiedliche Vorschläge für die Wahl der Spannungs- und Flußvariablen

für die Beschreibung thermodynamisch-verfahrenstechnischer oder hydrodynamischer Prozesse mittels Bindungsdiagrammen gemacht (Tabelle 1).

Tabelle 1. Vorgeschlagene Paare von Spannungs- und Flußvariablen für thermodynamisch-verfahrenstechnische oder hydraulische Prozesse

Spannungsvariable	Flußvariable	Anwendungsgebiet	Literaturstellen
Druck(differenz)	Volumenstrom	Hydraulik	/3/, /7/
Temperatur Temperatur	Wärmestrom Entropiestrom	reiner Wärmeübergang (statisch)	/1/ /3/
Temperatur Druck Temperatur	Entropiestrom Volumenstrom Entropiestrom	Thermodynamik	/3/ /7/
Chemisches Potential (einer Komponente)	Molstrom (einer Komponente)	Chemische Systemteile	/7/, /8/

Die Beschreibung hydraulischer Prozesse mittels der in Tab. 1 angegebenen Variablen beruht auf der hier häufig zulässigen Annahme, daß die Energieströme im Prozeß hinreichend durch das Produkt aus Druck und Volumenstrom dargestellt werden. Die Verwendung von Temperatur T und Wärmestrom dq/dt als Spannungs- und Flußvariable ist von Paynter für Prozesse mit reinem Wärmeübergang vorgeschlagen worden /1/. Später wurde gezeigt, daß man den Prozeß des statischen Wärmeübergangs zwischen zwei Systemen, die sich beide in einem Gleichgewichtszustand befinden, mittels der Variablen Temperatur T und Entropiestrom ds/dt beschreiben kann /3/. Unter den angegebenen Voraussetzungen gilt nämlich für die übertragene Wärmemenge dq = ds/T, wenn ds die übertragene Entropiemenge je Masseneinheit ist.

Nicht selten wird in der Literatur behauptet, daß man thermodynamische Prozesse (ohne chemische Reaktionen) grundsätzlich mittels der Spannungs- und Flußvariablen T und ds/dt beschreiben könne, z.B. /7/. Das dies nicht zutrifft, geht bereits aus /3/ S. 351 ff. hervor. Dort werden unter den einschränkenden Annahmen, daß der thermodynamische Prozeß (ohne chemische Reaktionen) geschlossen ist und sich in einem lokalen Pseudogleichgewicht befindet (daß also der Zustand des Prozesses sich so langsam ändert, daß die bekannten Beziehungen der "Thermostatik" gelten), die "konstitutiven Beziehungen" für eine reine Substanz durch ein C-Feld gemäß Bild 2 dargestellt. *Bemerkenswert ist, daß hierbei innerhalb der thermodynamischen* Energieregion zwei Arten von Spannungs- und zwei Arten von Flußvariablen

benutzt werden. In Erweiterung dieser Vorstellungen führte Karnopp dann
später "Pseudo-Bindungsdiagramme" ein, in denen durchweg zwei Paare von
Spannungs- und Flußvariablen, also insgesamt vier Variable, benutzt werden
/9/.

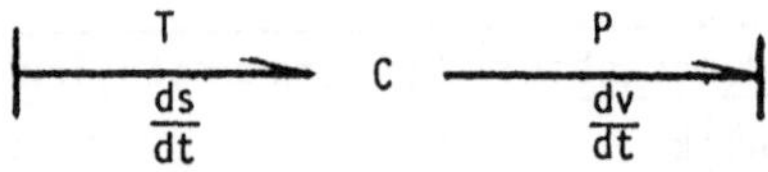

Bild 2. C-Feld zur Darstellung der
konstitutiven Beziehungen einer
reinen Substanz (/3/, S. 353).

Zweck der Anwendung von Bindungsdiagrammen
ist die Beschreibung dynamischer Prozesse,
d.h. in den hier betrachteten Fällen muß
die irreversible Thermodynamik zugrunde
gelegt werden. Obwohl die Entropie eigentlich auch hier für unentbehrlich gehalten
werden kann, da man durch sie den zweiten
Hauptsatz der Thermodynamik ausdrückt, ist sie für ("dynamische") Vorgänge
nicht definiert. Ihre Existenz wird allenfalls vorausgesetzt /10/. Die Ver-
fahren, eine in geeigneter Weise definierte Entropie unter der einschränkenden Voraussetzung, daß lokales Gleichgewicht vorliegt, für Nichtgleich-
gewichts-Prozesse zu bestimmen, sind obendrein kompliziert /11/.

3. Transformation von Zustandsdifferentialgleichungen in Bindungsdiagramme

Differentialgleichungen (oder entsprechende Integralgleichungen) mit der Zeit
als unabhängiger Variabler und Bindungsdiagramme sind gleichwertige Beschreibungen von dynamischen Prozessen, jedoch hat eine Beschreibung durch Bindungsdiagramme die im Abschn. 1 genannten Vorzüge.

Differentialgleichungen zur Beschreibung dynamischer thermodynamisch-verfahrenstechnischer Prozesse sind durchaus bekannt. Sie werden in der Regel
nicht über Bindungsdiagramme, sondern auf andere Weise (z.B. unmittelbar
aus Massen- und Engergiebilanzen) entwickelt. Man kann also den umgekehrten
Weg gehen, als den, der bisher bei der Anwendung von Bindungsdiagrammen
beschritten wurde, nämlich aus bekannten Differentialgleichungen die zuge-
hörigen Bindungsdiagramme entwickeln. So erhält man im Rahmen der Definitionen der Bindungsdiagramme Aufschluß über die Struktur solcher Prozesse.
Gleichzeitig ergeben sich, wie sich unten zeigen wird, nahezu zwangsläufig
diejenigen Spannungs- und Flußvariablen, mit denen überhaupt eine Beschreibung des Prozesses durch ein Bindungsdiagramm möglich ist.

Die hier zur Umformung gegebener Differentialgleichungen in Bindungsdiagramme angegebene Methode geht von einer mathematischen Beschreibung der
für typisierte Substrukturen von Bindungsdiagrammen dargestellten Zusammen-
hänge in ihrer allgemeinsten Form aus. Durch Vergleich dieser allgemeinen
mathematischen Beschreibungen mit den Zustandsdifferentialgleichungen des
betrachteten Prozesses kann man dessen Bindungsdiagramm entwickeln, wie
in Abschn. 4 am Beispiel eines nichtlinearen Prozesses gezeigt wird. Für die Behandlung dieses Beispieles genügt es, die beiden Substrukturen nach Bild 3 zu verwenden.

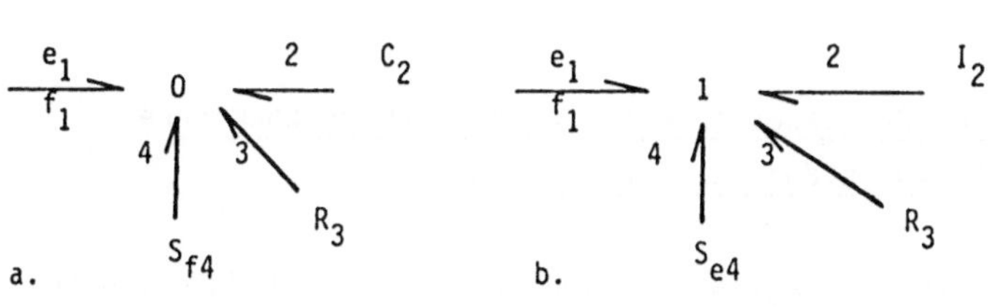

Bild 3. Zwei Substrukturen von Bindungsdiagrammen

Substruktur a wird allgemein durch die mathematischen Beziehungen beschrieben:

$$\frac{de_1}{dt} = \frac{b_1\, f_1(t)}{b_2\, h_1(e_1)} + \frac{b_3\, h_2(e_1)}{b_2\, h_1(e_1)} + \frac{b_4\, F_4(t)}{b_2\, h_1(e_1)} \tag{1}$$

$$\text{mit}\quad h_1(e_1) = -\left[\frac{d\phi_{c2}(e_2)}{de_2}\right]_{e_2=e_1} \quad,\quad h_2(e_1) = \left[\phi_{R3}^{-1}(e_3)\right]_{e_3=e_1} \quad,\quad \text{(1a,b)}$$

wobei ϕ_{R3}^{-1} die Umkehrfunktion von ϕ_{R3} ist.

Substruktur b wird beschrieben durch

$$\frac{df_1}{dt} = \frac{b_1'\, e_1(t)}{b_2'\, m_1(f_1)} + \frac{b_3'\, m_2(f_1)}{b_2'\, m_1(f_1)} + \frac{b_4'\, E_4'(t)}{b_2'\, m_1(f_1)} \tag{2}$$

$$\text{mit}\quad m_1(f_1) = -\left[\frac{d\phi_{I2}'(f_2')}{df_2'}\right]_{f_2'=f_1} \quad,\quad m_2(f_1) = \left[\phi_{R3}'(f_3)\right]_{f_3'=f_1} \quad \text{(2a,b)}$$

In diesen Beziehungen sind also verschiedene noch frei wählbare Funktionen enthalten. Diese sind so zu bestimmen, daß die Zustandsdifferentialgleichungen mit den Differentialgleichungen der Substrukturen identisch werden. Die Spannungs- und Flußvariablen dürften sich, wie das im unten behandelten Beispiel der Fall ist, meistens zwangsläufig ergeben. Das gesuchte Bindungsdiagramm kann man aus den Bindungsdiagrammen passend gewählter Substrukturen zusammensetzen.

4. Bindungsdiagramm eines Prozesses in einem durchströmten chemischen Rührkesselreaktor mit vollständiger Durchmischung

In einem von den Substanzen A_1, A_2, A_3 kontinuierlich durchströmten, gekühlten Rührkesselreaktor finde die exotherme chemische Reaktion $A_1 \longrightarrow A_2$ statt. Im Reaktor herrsche vollständige Vermischung der Substanzen. Eine bekannte mathematische Beschreibung dieses Prozesses ist

$$\frac{dc_1}{dt} = -r_1 + \frac{c_{1e} - c_1}{\theta} , \tag{3}$$

$$\frac{dT}{dt} = J_1 r_1 + \frac{T_e - T}{\theta} - h (T - T_k) . \tag{4}$$

Die Konzentration c_1 von A_1 sowie die Temperatur T innerhalb des Reaktors können als Zustandsvariable aufgefaßt werden; c_{1e}, T_e und T_k sind Komponenten des Eingangsvektors. Die Reaktionsgeschwindigkeit r_1 hängt von c_1 und T ab, $r_1 = r_1 (c_1, T)$. Bekannt ist der Ansatz von Arrhenius

$$r_1(c_1, T) = A c_1 e^{-E/RT} , \tag{5}$$

mit dem die Dgl. (3), (4) nichtlinear werden. Führt man durch

$$c_1 = a x_1 , \quad T = g (x_2), \quad c_{1e} = a u_1 , \quad \frac{T_e}{\theta} + h T_k = \tilde{g} (u_2)$$

die neuen Variablen $x_1(t)$, $x_2(t)$, $u_1(t)$, $u_2(t)$ ein, wobei zunächst a eine beliebige Konstante, $g(x_2)$ irgendeine differenzierbare Funktion von x_2, $\tilde{g}(u_2)$ irgendeine Funktion von u_2 ist, so erhält man aus (3)-(5):

$$\frac{dx_1}{dt} = \frac{u_1}{\theta} - x_1 \left[\frac{1}{\theta} + A\, e^{-E/R\ g(x_2)} \right] , \tag{6}$$

$$\frac{dg}{dx_2} \frac{dx_2}{dt} = J_1\, a\, A\, x_1\, e^{-E/R\ g(x_2)} - \left[\frac{1}{\theta} + h \right] g\,(x_2) + \tilde{g}\,(u_2). \tag{7}$$

Im ersten Schritt sollen die beliebigen Funktionen in Gl. (2) so bestimmt werden, daß Gl. (2) und (6) identisch werden. u_1 ist dabei mittels einer Quelle darzustellen. Man erkennt außerdem, daß $f_1 = x_1$ sein muß. So erhält man mit $b_1' = b_2' = -1$, $b_4' = 1$:

$$m_1\,(f_1) = -\,1/f_1 \quad , \quad m_2\,(f_1) = 0 \quad , \tag{8a-d}$$

$$e_1(t) = \frac{1}{\theta} + A\, e^{-E/R\ g(x_2)} \quad , \quad E_4'(t) = \frac{u_1}{\theta f_1} \ .$$

Aus Gl. (8c) kann man nun de_1/dt berechnen. Verwendet man dies, so läßt sich ableiten, daß Gl. (1) und Gl. (7) identisch sind für $b_1 = b_4 = 1$, $b_2 = b_3 = -1$, sofern

$$h_1(e_1) = -\frac{R}{J_1\, a\, E} \frac{g^2(x_2)}{(e_1 - \frac{1}{\theta})^2} \ , \quad h_2(e_1) = \frac{h + \frac{1}{\theta}}{J_1\, a} \frac{g(x_2)}{e_1 - \frac{1}{\theta}} \quad , \tag{9a-c}$$

$$F_4(t) = \frac{1}{J_1\, a} \frac{\tilde{g}\,(u_2)}{e_1 - \frac{1}{\theta}} \ .$$

Mit Gl. (8c) folgt aus den Gln. (9a) und (9b):

$$h_1(e_1) = -\frac{E}{J_1 aR} \frac{1}{(e_1 - \frac{1}{\theta})^2 \ln^2 \left[(e_1/A) - (1/A\theta) \right]} \ , \tag{10}$$

$$h_2(e_1) = -\frac{E}{J_1 aR} \frac{h + (1/\theta)}{(e_1 - (1/\theta))\, \ln \left[(e_1/A) - (1/A\theta) \right]} \ . \tag{11}$$

Aus den Gln. (2a) und (8a) bzw. (2b) und (8b) folgt unter Beachtung der allgemeinen Definition eines I-Speichers

$$\phi'_{I2}(f'_2) = \ln f'_2 + d_1 = \int_0^t e'_2(t)dt \ , \tag{12}$$

$$\phi'_{R3}(f'_3) = 0 \ , \tag{13}$$

mit beliebigem d_1. Schließlich folgt aus den Gln. (1a), (10) durch Integration

$$\phi_{c2}(e_2) = \frac{E}{J_1 aR}\left[\frac{-1}{(e_2-\frac{1}{\theta})\ \ln\ (\frac{e_2}{A} - \frac{1}{A\theta})} - \frac{1}{A}\int \frac{d\frac{1}{(e_2/A)-(1/A\theta)}}{\ln\frac{1}{(E_2/A)-(1/A\theta)}}\right]+d_2$$

$$= \int_0^t f_2(t)\ dt \ , \tag{14}$$

wobei das verbleibende Integral nicht durch elementare Funktionen dargestellt werden kann. Aus Gl. (11) erhält man schließlich die konstitutive Beziehung für das Element R_3:

$$f_3 = -\ \frac{E}{J_1 aR}\ \frac{h + \frac{1}{\theta}}{(e_3-(1/\theta))\ \ln\ [(e_3/A)-(1/A\theta)]} \ . \tag{15}$$

Das gewünschte Bindungsdiagramm ergibt sich hier, indem man die beiden Substrukturen nach Bild 3a,b unter Berücksichtigung der gemeinsamen Variablen e_1, f_1 zusammensetzt (Bild 4). Die Spannungsquellen S'_{e4}, S_{f4} sind gemäß Gln. (8d), (9c) "gesteuerten Quellen". Die Kausalitäten sind in Bild 4 entsprechend den für Bindungsdiagramme geltenden Regeln (/2/, /3/) eingetragen. Die Flußvariablen in diesem Bindungsdiagramm haben die Dimension einer Konzentration, sofern man a als dimensionslos definiert,

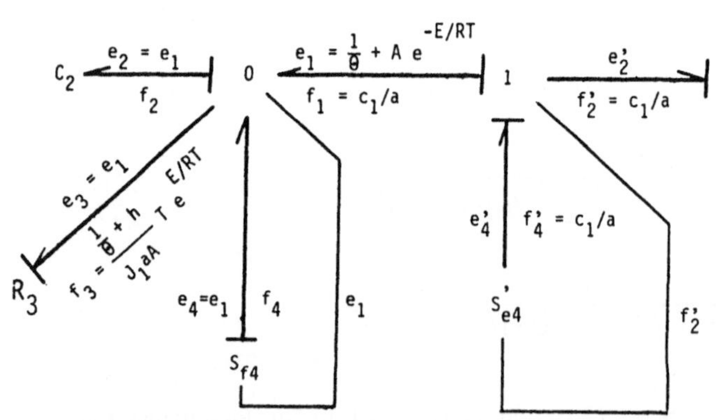

Bild 4. Bindungsdiagramm eines dynamischen chemischen Prozesses nach den Gln. (3) bis (5).

die Spannungsvariablen die Dimension s^{-1}. Sofern man für die Bestimmung von f_2 die konstitutive Beziehung für die 0-Verzweigung, für die Bestimmung von e_2' die konstitutive Beziehung für die 1-Verzweigung benutzt und beachtet, daß $f_4 = F_4(t)$, $e_4' = E_4'(t)$, kann man aus den Gln. (12), (14) zwei Integralgleichungen mit den Zustandsvariablen c_1, T aufstellen, die an die Stelle der Gln. (3), (4) treten.

5. Schlußbemerkungen

Das Bindungsdiagramm nach Bild 4 hat sich quasi zwangsläufig aus den vorgegebenen Dgln. (3), (4) ergeben. Für eine genaue Analyse dieses Bindungsdiagramms ist hier kein Platz. Einfacher ist die Analyse eines Bindungsdiagramms, das man erhält, wenn man als Näherung für r_1 einen linearen Ansatz macht.

Eine Erweiterung des angegebenen Verfahrens auf Systeme von Zustandsdifferentialgleichungen mit mehr als zwei Zustandsvariablen ist dadurch möglich, daß man auch Substrukturen benutzt, in denen mindestens zwei Bindungen jeweils an einem Ende noch nicht durch ein bestimmtes Element belegt sind. Statt der Gl. (1) oder (2) verwendet man dann also z.B. eine Dgl., in der die Variablen e_1, f_1, e_2, f_2 enthalten sind.

Formelzeichen

A	Reaktionsgeschwindigkeitskonstante
A_λ	Bruttoformel der chemischen Verbindung λ, $\lambda = 1,2,3$
a	Konstante mit beliebigem Wert
b_i	bezeichnet die positive Richtung des Energiestromes der Bindung i
c_1	Konzentration der Verbindung A_1 im Reaktor
c_{1e}	Konzentration von A_1 im Zulauf
d_1, d_2	Integrationskonstanten
E	Aktivierungsenergie
$E_i(t)$	von der Quelle $S_{e\,i}$ erzeugte Spannungsvariable
$e_i(t)$	Spannungsvariable der Bindung i
$F_i(t)$	von der Quelle $S_{f\,i}$ erzeugte Flußvariable
$f_i'(t)$	Flußvariable der Bindung i

$g, \tilde{g}$ beliebige Funktionen

h Parameter

h_1, h_2 Funktionen einer Spannungsvariablen

i Nummer einer Bindung

J_1 Parameter

m_1, m_2 Funktionen einer Flußvariablen

R Gaskonstante

r_1 Reaktionsgeschwindigkeit bezüglich A_1

T Temperatur

T_e Zulauftemperatur

T_k Temperatur des Kühlmittels

u_1, u_2 (unabhängige) Variable

x_1, x_2 (abhängige) Variable

θ Mittlere Verweilzeit im Rührkesselreaktor

$\phi_{R\nu}$ Funktion zur Definition des νten R-Elementes

$\phi_{C\nu}$ Funktion zur Definition des νten C-Elementes

$\phi_{I\nu}$ Funktion zur Definition des νten I-Elementes

Literatur

/1/   Paynter, H.M.: Analysis and Design of Engineering Systems. Cambridge
       (Mass.). The M.I.T. Press, 1960 und 1961.

/2/   Karnopp, D., u. R.C. Rosenberg: Analysis and Simulation of Multiport
       Systems. The M.I.T. Press, Cambridge (Mass.), London 1968.

/3/   Karnopp, D., u. R.C. Rosenberg: Systems Dynamics: A Unified Approach.
       John Wiley & Sons, New York, London, Sydney, Toronto 1975.

/4/   Schöne, A.: Systembeschreibung durch simultane Diagramme. Regelungs-
       technik 24 (1976) H. 2, S. 37-42, H. 3, S. 80-85.

/5/   Karnopp, D.: Die ENPORT-Simulationsprogramme. In: Schöne, A.: Simu-
       lation technischer Systeme. Band 1: Grundlagen der Simulationstechnik.
       Carl Hanser Verlag, München, Wien 1974. S. 93-102.

/6/   Altmann, A.: Interdisziplinäre Systemanalyse: Eine Strukturalgebra
       der Bonddiagramme. Springer-Verlag, 1982.

/7/   Mansour, M., u. A. Altmann: Modellbildung dynamischer Systeme. 3.
       Symposium Simulationstechnik 1985. Informatik-Fachberichte 109.
       Springer-Verlag Berlin u.a.O. 1985. S. 37-49.

/8/   Oster, G.F., u. D.M. Auslander: Topological Representation of
      Thermodynamic Systems. I. Basic Concepts. II. Some Elemental
      Subunits for Irreversible Thermodynamics. Journal of the Franklin
      Institute, Vol. 292 (1971), S. 1-17, 77-92.

/9/   Karnopp, D.: State Variables and Pseudo Band Graphs for Compressible
      Thermofluid Systems. Transactions of the ASME. Journal of Dynamic
      Systems, Measurement, and Control Vol. 201 (1979), S. 201-204.

/10/ Meixner, J.: Thermodynamik der Vorgänge in einfachen fluiden Medien
      und Charakterisierung der Thermodynamik irreversibler Prozesse.
      Zeitschrift für Physik (A) Bd. 219 (1969), S. 79-104.

/11/ Glansdorff, P., u. I. Prigogine: Thermodynamic Theory of Structure,
      Stability and Fluctuations. Wiley-Interscience, London u.a.O. 1971.
      2. Nachdruck 1977.

Biological and Physiological Systems

The use of Artificial Intelligence for Simulation of Metabolic
Processes
J.R. Reichl

Population Dynamics of Daphnia Magna: Simulations using the
Individuals Approach
V. Fitsch, H. Kaiser

Estimation of Individual Growth Curves from Aggregate Data
W. Wosniok

Sleeping Stem Cells: A Model of OF Stem Cells under Continuous
Stress
G. Pabst

A Mathematical Method of Modelling and Simulating Biological
Structure Control Systems
T. Vogelsaenger

Simulation in Electrophysiological Pharmacology: Specific
Interactions of Antiarrhythmic Agents with Ion Channels of
the Cardiac Cell Membrane
D. Hafner, F. Berger, U. Borchard

Simulation of the Human Blood Circulatory System with the
help of an Uncontrolled Pulsatile Model and its Validation
T. Sikora, D.P.F. Möller, V. Pohl, E. Hennig

Model-Reduction for the Parameter Identification of an
Uncontrolled Pulsatile Model of the Cardiovascular System
V. Pohl, D.P.F. Möller, T. Sikora, E. Hennig

On the improved Estimation of the Compliance Parameters of
the Physiologically Closed Cardiovascular System
A. Tanha, H. Maftoon, G. Thiele, D.P.F. Möller, D. Popovic

Some aspects of the application of Neurodynamical Models for
the Simulation of Central Regulation and Dysregulation
O. Hoffmann

Measuring Symptoms in Parkinson's Disease with a Tracking
Device
S.S. Hacisalihzade, C. Albani, M.A. Mueller

The use of Artificial Intelligence for Simulation of Metabolic Processes

Jan R. R e i c h l, Hohenheim

Zusammenfassung. Die Arbeit beschreibt ein System zur Simulation der Stoffwechselprozesse zusammen mit den strategischen Regeln, die einer prozeduralen Produktionsmethode der künstlichen Intelligenz entsprechen. Der Simulationsteil benutzt algebraischen- und Differentialgleichungen. Der strategische Teil besteht aus Kontext, Produktion und Interpreter. Die höhere strategische Ebene benutzt folgende 6 Prinzipien zur Konstruktion von Hypothesen: Ähnlichkeit, Polarität, Verschiedenartigkeit, Symmetrie, Erregbarkeit und Rhythmizität.

Summary. A system for simulation of metabolic processes together with strategy rules, which correspond to the procedural production method of the artificial intelligence are presented. The simulation part uses algebraical and differential equations. The strategy part consists of the Context, Production and Interpreter. The higher strategy level uses the following 6 principles for construction of hypotheses: Similarity, Polarity, Diversity, Symmetry, Excitability, and Rhythmicity.

1. Introduction

The artificial intelligence (AI) methods can be divided into d e c l a r a t i v e (propositional logic, predicate calculus), and p r o c e d u r a l (knowable behavior, causal chain construction). There is amazing similarity between the procedural production methods, and the simulation techniques, since both can use the network flow diagrams, and can be formulated by similar type of equations, or strategy rules. This similarity provokes a question to what extent are AI methods artificial, and on the other hand, to what extent can simulation techniques be intelligent.

In the early machine AI the use of the technique of decission trees and computer algorithms lead to a serious problem of exponential or combinatorial explosion. According to DEHN and SHANK (1982), the AI is now turning more toward attempted simulation of human heuristic methods, which differ from algorithms in that they do not specify how to solve an entire problem, and do not examine all possible solutions, but suggest a way how to find a

most intelligent solution. According to these authors, research
in AI showed, that understanding is basic to intelligence and
that intelligence is understanding things in terms of what one
already knows.
The understanding process, which doesn't need to be conscious,
assumes, that at least two structures match together. In biolo-
logical systems the coding of genetic information represent an
example of this match. The protein structure which is produced
in this way, contains a long-term information, which is used to
carry out a specific metabolic process. The conformation of the
tertiary structure of an enzyme exhibits an adaptable short-term
memory (KATZ and WESTLEY,1979). The neural system represents an
additional, global memory system.
The metabolic process in cells is a combination of at least two
sequences of transformations: (a) synthesis and breakdown of
small molecules (nutrients,metabolites,products), which correspond to the energy metabolism, and (b) synthesis and breakdown
of macromolecules (enzymes,coenzymes,receptors), which contribute mainly to the protein metabolism. However, the energy metabolism is carried out on the surface of the macromolecules,so that
the processes (a) and (b) depend each on the other. Crucial for
these transformations is the exactness of the sequence of reactions. This gives us a possibility to describe the metabolic processes not only by means of mass action (using differential equations), but also by means of sequential transformations (using
algebraical equations). Once the system is adapted to a specific
situation, the short-term memory of the system functions analogous to the 'reasoning backwards' in the AI, i.e. the system is
goal-driven. This process represents a rote learning. Physically
such an adapted system correspond to a network of fully occupied
surfaces or spaces. In this case the emptying of the last one
controls the whole flow. In an nonadapted system,i.e. in a tran-
sient state, the control impulses are coming predominantly from
the inputs.

2. **The Simulation System**
The presented simulation system, despite its predictive value,
is at first instance an interpretative system, constructed, iden-

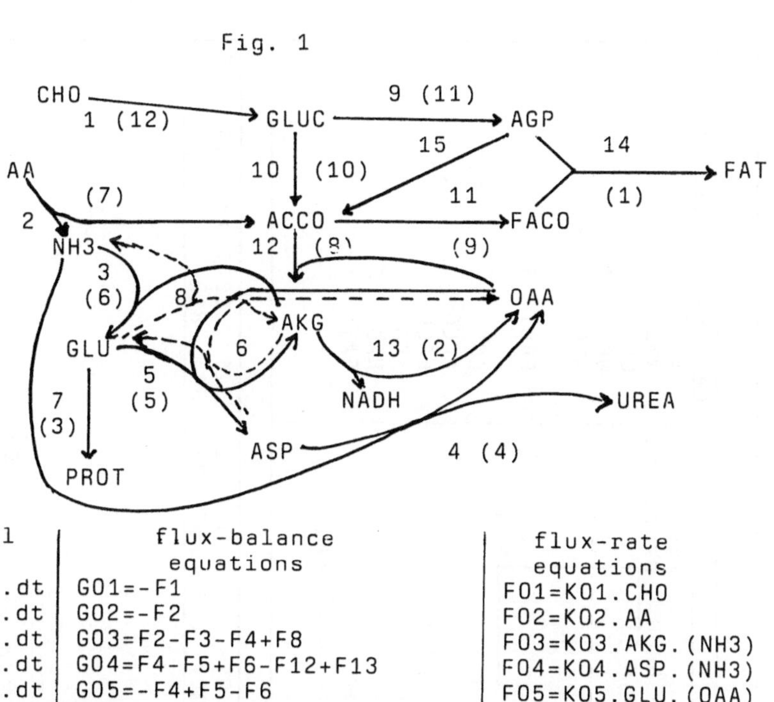

differential equations	flux-balance equations	flux-rate equations
CHO =CHO +G01.dt	G01=-F1	F01=K01.CHO
AA =AA +G02.dt	G02=-F2	F02=K02.AA
NH3 =NH3 +G03.dt	G03=F2-F3-F4+F8	F03=K03.AKG.(NH3)
OAA =OAA +G04.dt	G04=F4-F5+F6-F12+F13	F04=K04.ASP.(NH3)
ASP =ASP +G05.dt	G05=-F4+F5-F6	F05=K05.GLU.(OAA)
GLU =GLU +G06.dt	G06=F3-F5+F6-F7-F8	F06=K06.ASP.(AKG)
GLUC=GLUC+G07.dt	G07=F1-F9-F10	F07=K07.GLU
ACCO=ACCO+G08.dt	G08=F2+F10-F11-F12+F15	F08=K08.GLU
AKG =AKG +G09.dt	G09=-F3+F5-F6+F8+F12-F13	F09=K09.GLUC
FACO=FACO+G10.dt	G10=F11-F14	F10=K10.GLUC
AGP =AGP +G11.dt	G11=F9-F14-F15	F11=K11.ACCO
PROT=PROT+G12.dt	G12=F7	F12=K12.ACCO
UREA=UREA+G13.dt	G13=F4	F13=K13.AKG
FAT =FAT +G14.dt	G14=F14	F14=K14.FACO.(AGP)
NADH=NADH+G15.dt	G15=F1-F3+F8-F9+F10-F11+.. ..F12+F13+F15	F15=K15.AGP

tified, and validated by the following stepwise procedure:
(a) the system of equations is written using the stored elements
according to the given goals, either in an input-driven, or a
goal-driven sequence, (b) the fluxes of individual reactions are
balanced and their size determined by a stepwise sequential procedure, (c) using the determined fluxes, the rate constants for
individual reactions are calculated, (d) by repeated dynamical
simulation the initial concentrations of chemicals are adjusted,
(e) validation of the system occur at the end of simulation by
comparing the simulation results with all available data on all
levels of the system.

Table 1

			input-driven sequence of equations	recycling procedure	goal-driven sequence	
CHO	1	branching ratio	– CHO + 1.8 GLUC		GLUC	12
AA	2	ratio	– AA + ACCO + 1.2 NH3 + 2 NADH		NH3	7
NH3	3	R3	– NH3 – AKG + GLU – NAD(P)H	←	GLU	6
NH3	4	R4	– NH3 – ASP + OAA + UREA		UREA	4
OAA	5		– OAA – GLU + ASP + AKG		ASP	5
ASP	6		– ASP – AKG + GLU + OAA		–	
GLU	7	R7	– GLU + PROT		PROT	3
GLU	8	R8	– GLU + AKG + NH3 + NADH ——— NH3		–	
GLUC	9	R9	– GLUC + 2 AGP – 2 NAD(P)H		AGP	11
GLUC	10	R10	– GLUC + 2 ACCO + 4 NADH		ACCO	10
ACCO	11	R11	– ACCO + .14 FACO – 1.8 NAD(P)H	←	FACO	9
ACCO	12	R12	– ACCO – OAA + AKG + NAD(P)H		AKG	8
AKG	13		– AKG + OAA + 3 NADH		NADH	2
FACO	14		– FACO – .33 AGP + .33 FAT		FAT	1
AGP	15		– AGP + ACCO + 3 NADH ——— ACCO		–	

Table 2

Gi	Fi	01	02	03 R03	04 R04	05	06	07 R07	08 R08	09 R09	10 R10	11 R11	12 R12	13	14	15
01 CHO		-162														
02 AA			-115													
03 NH3			1.2	-1	-1				1							
04 OAA					1	-1	1						-1	1		
05 ASP					-1	1	-1									
06 GLU				1		-1	1	-1	-1							
07 GLUC		1.8								-1	-1					
08 ACCO			1								2	-1	-1			1
09 AKG				-1		1	-1		1				1	-1		
10 FACO												.14			-1	
11 AGP										2					-.33	-1
12 PROT								1								
13 UREA					1											
14 FAT															.33	
15 NADH		2		-1					1	-2	4	-1.8	1	3		3

2.1 The System of Equations

The flow diagram of Fig. 1 represents a very simplified example
of a network for energy metabolism. In this example the enzymes,
coenzymes, the intermediary steps, and the compartmentation between the particular organs or organelles are omitted. The input
variables are the carbohydrates (CHO) and amino acids (AA), the
output variables are protein, urea, fat, and NADH. In Fig. 1 the
differential equations (CHO,AA,etc.), flux-balance equations(Gi)
and flux-rate equations(Fi) are also presented. They are ordered
by an input-driven sequence. Within the dynamical simulation all
three types of equations are used, for the balance of fluxes
only the equations Gi.

Tab. 1 shows the source equations of this example in an algebraical form and Tab. 2 in a matrix form. They are ordered in an
input-driven sequence. On the right margin of Tab. 1 the goal-
driven sequence of these equations is also shown. In this system
the minus sign means, that the particular chemical occur at the
left-hand side (LHS) of a reaction, the plus sign indicates the
right-hand side (RHS). In Tab. 2 the columns (Fi) indicate the
individual reactions, the rows (Gi) the particular chemicals.
The numbers in the table are stoichiometries, given in moles for
the intermediary chemicals and in grams for the input chemicals.

2.2. The Solution Technique

Within the dynamical simulation, the sequence of reactions
is insignificant, since their solutions occur simultaneously
in small integration intervals. On the other hand, the sequence
of reactions is essential for the algebraical balance of fluxes,
since the solutions occur subsequently on the diagonal, where
the corresponding chemical is set to zero. The stoichiometrical
amount of the corresponding transferred chemical indicates the
size of the particular flux. The way by which the stepwise solution is carried out, is evident from Tab. 1 and Tab. 2. In This
example 3 chemicals occur above the diagonal (NH_3,OAA and AGP).
These chemicals can be balanced to zero only by repeated calcu-
lations. This is done in two cycles. The cycle for balancing of
NH_3+OAA is placed between the reactions 3 and 8, the cycle for
balancing ACCO+OAA between reactions 11 and 15. In this way the

branching ratio between reactions 3 and 4 (R3/R4) and between 11
and 12 (R11/R12) is solved. In contrast to this input-driven sequence,the goal-driven sequence arrives to the solution directly
without cycling and it occurs naturally in an adapted living system. On the other hand, the input-driven solution is typical for
systems which are in transition between two states. The dynamic
simulations can demonstrate, that the adapted state corresponds
to a dynamic steady-state in that the inputs are equal to the
outputs, and the initial and final concentrations of the intermediary chemicals are equal to each other. If the system is not
continuously but regularly fed, the concentrations of chemicals
change in the course of time, but despite of it, they still are
in a steady-state. The dynamical steady-state is the most relia-
ble criterium for the stability of the simulated system.

3. The Strategy Rules

According to the HANDBOOK OF ARTIFICIAL INTELLIGENCE (1983),
early attempts to develop intelligence by simulation of random
mutation and natural selection failed. Later, the researchers in
AI adapted the view that a learning system cannot gain knowledge
by starting 'from scratch', that is without any knowledge
at all. Some knowledge must be employed by every learning system
to understand the information provided by the environment. The
knowledge can then grow and develop to a higher level. If the
knowledge is general enough to produce details, the details may
be forgotten, i.e. scratched off from the memory.
This system uses two strategy levels. The lower strategy level
correspond to the elementary simulation structure described in
the preceding paragraph in that the use of repeated simulations
adjust the branching points and the steady-state concentrations.
This lower level of simulation together with the lower level of
strategy rules correspond to the rote learning,which is just me-
morization. Higher strategy level uses higher forms of learning
and has the purpose of finding explicite rules which can replace
the adjusting or adapting procedures.

3.1 Lower Strategy Level

Using the terminology of AI, the production system consists of

three parts: (a) C o n t e x t, which represents information
about the circumstances surrounding an act or event (variables,
reactions,parameters), (b) P r o d u c t i o n, which represents
condition-action sequence (sequence of equations,IF-THEN rules),
(c) I n t e r p r e t e r, which explains the meaning, transla-
lates, or directs (management,interaction with the user).

3.1.1. C o n t e x t

In the presented simulation system the Context is organized into
three types of string variables: (a) the elementary string sto-
res the original data sets in the same order as they occur in an
living system, for example in a protein following a sequence of
amino acids, in a pathway following a sequence of metabolites,or
in a feed some natural order of nutrients; (b) adapted string
is a product of Production and will be defined in the subsequent
paragraph; (c) semantic string contains full names or sentences
corresponding to the elements stored in the elementary or adapted strings, it is used by the Interpreter. Construction of these strings correspond to a static simulation of the structure of
a living system.

3.1.2. P r o d u c t i o n

This part of the system consists of rules, how to order from the
elements of the Context, the sequence of events, i.e. reactions,
according to the goals given by the Interpreter. The product of
this procedure is an adapted string and by the computer program
written simulation equations. This procedure can for example construct a folded protein by setting particular amino acids together for a particular function, or construct a set of metabolic
pathways for a particular tissue. Stored pathways can be used
either in full details, or be comprised just to the branching
points. This procedure simulates naturally occuring synthesis
and decay of the living structure.

3.1.3. I n t e r p r e t e r

The production procedure cannot carry out any task without kno-
wing the goals. The general goals are specified by the user, the
local goals, for example what is the net product of a particular
tissue, determines the simulation system itself. The Interpreter
directs the whole system and presents the results by use of the
semantic strings. It also directs selective rewriting of the

information from one string to another. This manipulation of
strings simulate the natural process of mapping informations in
the macromolecules and neural centers.

3.2 Higher Strategy Level

The lower level uses predominantly the elementary data base and
simulation is carried out by adjusting and adaptation procedures.
The rote learning of this level makes no hypotheses. The higher
strategy level uses principles, which provide a basis for making
hypotheses. At this level three forms of learning are possible:
(a) deductive hypothesizing of missing details, (b) inductive
hypothesizing of general rules, (c) hypothesizing of a whole
structure by using analogy. In this system six built-in principles are used. Despite of the fact, that all act together, there
is advantage in lining the first two to the Context, next two
to the Production, and the last two to the Interpreter.

3.2.1. Context Principles

The Context of the presented system contains information about
the elementary relations and reactions which were determined in
experiments. This part should hypothesize explicite rules for
data reduction and production.

(a) **Similarity** (headword: stoichiometry) is the most elementary
principle for data reduction. For example the heat production
can be calculated from the body weight $Q(Kcal) = 70 \ W^{.75}(Kg)$, or
from respiratory parameters $Q(Kcal) = 25.9 \ CO_2(M) + 86.3 \ O_2(M) -$
.8 urineN(g), or from the wind velocity and outside temperature.
Nutrients or cells can be classified by their products according
to specific environments and single reactions by their stoichiometries.

(b) **Polarity** (headword: turnover rate) describes the donor-
acceptor, or synthesis-decay relationships. For example the heat
production can be calculated by $dQ/dt = k \ A(T_R - T_E)$, where A is
the surface and $T_R - T_E$ the difference between rectal and environmental temperatures, or the growth of a substance can be described in terms of synthesis (k_s) and decay (k_d) by equations

$$dC/dt = k_s - k_d \ C, \quad or \quad C_t = (k_s/k_d)(1 - \exp(k_d t))$$

An example for explicite calculation of the rate constants and
of initial concentrations is presented in Fig. 2a. This system,

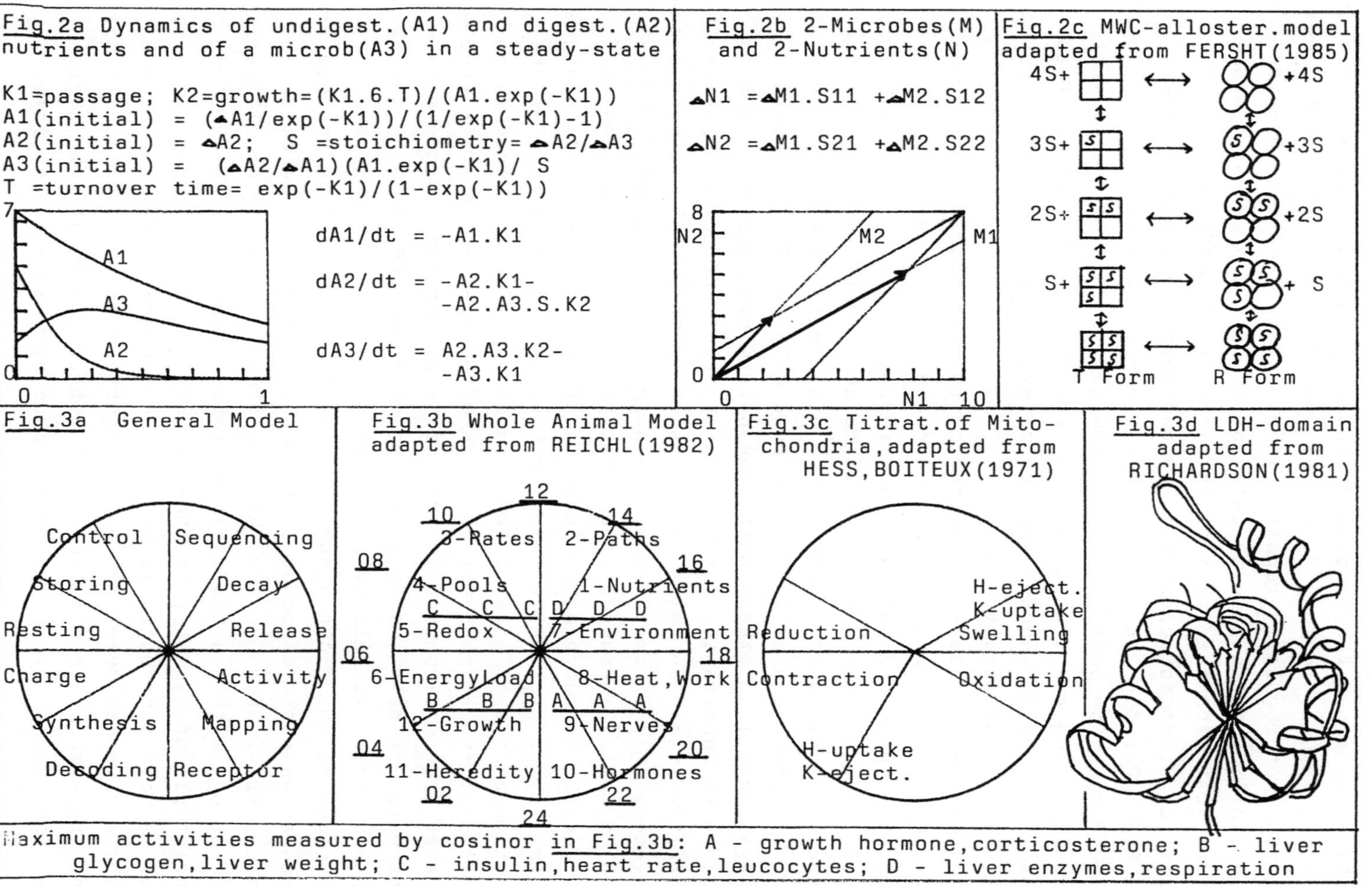

Fig.2a Dynamics of undigest.(A1) and digest.(A2) nutrients and of a microb(A3) in a steady-state
K1=passage; K2=growth=(K1.6.T)/(A1.exp(-K1))
A1(initial) = (▲A1/exp(-K1))/(1/exp(-K1)-1)
A2(initial) = ▲A2; S =stoichiometry= ▲A2/▲A3
A3(initial) = (▲A2/▲A1)(A1.exp(-K1)/ S
T =turnover time= exp(-K1)/(1-exp(-K1))
dA1/dt = -A1.K1
dA2/dt = -A2.K1- -A2.A3.S.K2
dA3/dt = A2.A3.K2- -A3.K1
7
A1
A3
A2
0
0
1
Fig.2b 2-Microbes(M) and 2-Nutrients(N)
▲N1 =▲M1.S11 +▲M2.S12
▲N2 =▲M1.S21 +▲M2.S22
8
N2
M2
M1
0
0
N1
10
Fig.2c MWC-alloster.model adapted from FERSHT(1985)
4S+
+4S
3S+
+3S
2S+
+2S
S+
+ S
T Form
R Form
Fig.3a General Model
Control
Sequencing
Storing
Decay
Resting
Release
Charge
Activity
Synthesis
Mapping
Decoding
Receptor
Fig.3b Whole Animal Model adapted from REICHL(1982)
12
10
14
3-Rates
2-Paths
08
16
4-Pools
1-Nutrients
C C C
D D D
5-Redox
7-Environment
06
18
6-EnergyLoad
8-Heat,Work
B B B
A A A
12-Growth
9-Nerves
04
20
11-Heredity
10-Hormones
02
22
24
Fig.3c Titrat.of Mito- chondria,adapted from HESS,BOITEUX(1971)
H-eject.
K-uptake
Swelling
Reduction
Oxidation
Contraction
H-uptake
K-eject.
Fig.3d LDH-domain adapted from RICHARDSON(1981)

Maximum activities measured by cosinor in Fig.3b: A - growth hormone,corticosterone; B - liver glycogen,liver weight; C - insulin,heart rate,leucocytes; D - liver enzymes,respiration

which describes the passage of nutrients and of a growing microb
from a vessel, is in a dynamical steady-state with a continuous
output but regularly repeated inputs. This condition provide the
possibility for explicite calculation of all necessary parameters just from the input of nutrients ($\triangle$A1, $\triangle$A2) and from the
rate constant for passage of an undigestible substance ($\triangle$A1).

3.2.2. Production Principles

The Production part of this system has to fulfill the task of
constructing complex structures from the elements of the data
base. This part should hypothesize explicite rules for construction of these structures.

(c) **Diversity** (headword: isotherm) is a principle, which describe events of competition and cooperation. In Fig. 2b the
parallelogram and two equations represent a system of two microbes and two nutrients. The balance determines the proportion of
these two microbes in the system,despite the fact, that they may
have different growth rates in pure culture. The explicite rule
says, that in a steady-state the determining factor is the balance of substrates, not the growth rate. Another example is the
NAD/NADH ratio in cells, which is the same for all dehydrogenase
systems in one compartment, for example in a cytosol.

(d) **Symmetry** (headword: coordination) is a principle, which
describe the coordinated function of single tissues in an organism, or of proteins within the tissues. The structure of the
active centre of alcohol dehydrogenase may be mentioned as an
example: the hydrophobic residues of neutral amino acids keep a
water free environment in the inside of the protein;one molecule
of water is held inside by the polar groups of zinc and serine;
NAD molecule is held by the polar group of threonine; and on the
surface of the protein one nitrogen of the histidine residue serves as an electron donor an the other nitrogen as an electron
acceptor. In Fig. 2c the MWC model for functioning of allosteric
reactions within a coordinated protein structure, which uses the
principle of symmetry,is shown.In Fig.3d the presented structure
of lactate dehydrogenase also shows a symmetrical structure of
the enzyme. However, two only slightly different configurations
of the same protein molecule may differ drastically in their
action, which may be the consequence of the next principle.

3.2.3. Interpreter Principles

The Interpreter of this system should manage the task of explaining the functioning of complex structures and hypothesize
explicite rules for the action of these structures.

(e) **Excitability** (headword:motor function) is a principle, which
serves as a base for the hypotheses about the origin of the
actions in living systems. The above mentioned NAD/NADH redox
ratio which can be calculated from different dehydrogenase systems, is controlled by the ATP/ADP ratio. The sequential order
of amino acids rather than the amino acid composition of a protein determines the pKa of all ionic groups in it; if ATP synthesis and hydrolysis move electrons along the protein chain,
cyclic work performance can occur (LING,1984). In this way not
only the metabolic but also the transport processes in the cells
may be hypothesized. Fig. 3a shows a model, which link the complex structure within the cells, or within the whole organism,
to the specific functioning of the single elements. In the spite
of this hypothesis the single backbones of the symmetrical structure of an enzyme, which is shown in Fig. 3d, may have diffe-
rentiated fine structure and provide different functions in relation to the other structures in their environment. The presented system possess an opportunity to construct such a structure
from single amino acids.

(f) **Rhythmicity** (headword: cycles) is a principle, which sup-
ports the excitability principle. As already mentioned, the
cyclic work performance in cells is linked to the ATP synthesis
and hydrolysis. According to HESS and BOITEUX (1971), glycolytic
oscillations are generated by feedback activation and inhibition
of the allosteric phosphofructokinase coplex by the adenosine
phosphates to the other kinases. The periodic changes of
the intermediates are coupled to the NAD/NADH system by two de-
hydrogenases in cytosol. Fig. 3c shows another example, which
demonstrates a cycling in mitochondria: concomitant with the os-
cilating ion movements and volume changes, periodic variations
of the redox state of respiratory carries can be recorded. Duration of the whole cycle is in the magnitude of seconds. This
example is in accordance with a part of the general model of the
Fig. 3a. In Fig. 3b a model for a whole animal,set in a frame of

a 24-hour cycle, is presented. The model elements are in good
agreement with some experimentally measured activities by the
method of cosinors.The submodels 1, 2, 7,and 8 of this model are
linked to the outer and inner surfaces of the whole animal(skin,
gastrointestinum)and decay functions, the submodels 3, 4, 9, and
10 cover the transport and metabolic functions inside and between the tissues controlled by hormones and nerves, and the submodels 5, 6, 11, and 12 correspond to the motor and synthetic
functions on the cellular level of an organism.

4. Conclusion

(a) There is no essential difference between the human, biological and artificial intelligence. All forms of intelligence can
be described by the same rules and follow the same principles.

(b) The most important contribution of the AI research is the
result, that intelligence cannot develop from a scratch by a
random process without interference from the outside.

(c) The metabolic processes in an organism follow principles,
which can be classified as intelligent. Simulation of biological
processes in an advanced form include also a simulation of biu-
logical intelligence.

(d) The use of AI for simulation of metabolic processes also
has to be intelligent. An intelligent solution does not mean
the most profitable solution with the lowest deviation, but one
which can interpret the deviations, the multiplicity and the
structure of the system.

References.
(1) DEHN,N.,SHANK,R.,1982: Artificial and human intelligence,in:
Handbook of Human Intelligence (ed.R.J.Sternberg), Cambr. Univ.
Press, 3-28; (2) FERSHT,A.,1985: Enzyme Structure and Mechanism,
2nd, Freeman; (3) Handbook of Artificial Intelligence, 1981:
vol.1 (ed. A.Barr,E.A.Feigenbaum), Pitman; (4) dtto, 1982 : vol.
2 (ed. A.Barr,E.A.Feigenbaum); (5) dtto, 1983: vol.3 (ed. P.R.
Cohen,E.A.Feigenbaum); (6) HESS,B.,BOITEUX,A., 1971: Oscillatory
phenomena in biochemistry, Ann.Rev.Biochem., 40, 237-258; (7)
KATZ,M.,WESTLEY,J., 1979: Enzymic Memory, J.Biol.Chem., 254,
9142-9147; (8) LING,G.N., 1984: In Search of the Physical Basis
of Life, Plenum; (9) REICHL,J.R., 1982: Computersimulation des
Protein- und Fettstoffwechsels bei Tieren,in: Simulationstechnik
(ed. M.Goller), Springer, 373-378; (10) RICHARDSON,J.,S., 1981,
The anatomy and taxonomy of protein structure, Adv.Prot.Chem.,
34, 167-339.

Population dynamics of Daphnia magna – Simulations using the individuals' approach

Vera Fitsch and Heinrich Kaiser, Aachen

Zusammenfassung. Ausgehend von individuellen Lebensdaten haben wir zunächst ein Modell des Lebensablaufs einzelner Daphnien entwickelt und anschließend mit Hilfe des Individuenansatzes eine *Daphnia magna* Population simuliert. Die Validität des Modells wurde durch Vergleich der simulierten Populationen mit entsprechend gehaltenen Laborpopulationen überprüft. Die gute Übereinstimmung zwischen simulierter und natürlicher Population spricht für die Richtigkeit des Populationsmodells und für die allgemeine Anwendbarkeit des Individuenansatzes.

Summary. By using a simulation approach with individual life table data, we have been able to predict the population dynamics of *Daphnia magna* under different environmental conditions. The simulated population dynamics proved to agree closely with the observed dynamics of natural populations grown under the corresponding environmental conditions. This indicates the validity of both the population model and the individuals' approach.

1. Introduction

One main subject of population research is the investigation of those causal relationships which determine the population processes. Normally the carrying capacity and intrinsic rate of natural increase are used to describe the population dynamics. However, proceeding in this way has some shortcomings: for example, it does not take into account the age structure of a population, which influences the subsequent development of a population. In order to obtain a more detailed insight into the causal mechanisms of the population dynamics of *Daphnia magna*, we have applied the individuals' approach (Kaiser (1975)to predict the population dynamics on the basis of the life histories of the individuals and of the natural variability between the animals.

In its natural environment *Daphnia magna* shows a typical population cycle during the course of the year. The population density, age distribution, and sex ratio are determined by environmental parameters, such as temperature, photoperiod and food abundance. In order to simplify the investigation of the causal relationships operating within the populations we reared laboratory populations of *Daphnia magna* under controlled constant conditions.

The typical result of such an experiment is shown in Fig. 2: starting with
5 newborn animals the population density increased up to a maximum before
decreasing and subsequently fluctuating around an equilibrium value (conditions: 25°C, LD 14:10, food = $5 \cdot 10^7$ Chlorella cells/d). Because the individuals' approach assumes that the population dynamics results from the characteristics of the individuals and is based on individual life histories,
it was necessary to investigate the complete life history of *Daphnia magna*.
For this reason, we reared single daphnids at different combinations of
temperature, photoperiod and food concentrations. Both types of experiment,
whether using individuals or using populations, were conducted under environmental conditions which are common in the field. The results of the experiments using single daphnids enabled us to establish the relationships between
the life table data and these three environmental factors (Fitsch & Ratte,
in prep.). Additionally we obtained the specific distribution of each life
history parameter. These distributions are necessary to describe the natural
variability among the individuals of a population. On the basis of this detailed life history information, we developed a model of a single daphnid's
life in order to simulate the dynamics of the population under different
environmental conditions.

2. The life cycle of a daphnid

When applying the individuals' approach, the first step in modelling a population is to describe the life cycle of a daphnid as completely as possible.
In the present context we ignore the possibility of sexual reproduction, because this plays only a minor role in the experiments. With parthenogenetic
reproduction the life of a daphnid used in the model can be described in
terms of:
1. Feeding(rate), growth(rate) and mortality(rate)
2. Development(rate) of embryos and juventils
3. Reproduction (number of parthenogenetically-produced eggs).

The newborn daphnid starts feeding on algae, grows, and develops until it
reaches the age of first reproduction. On becoming mature it may produce
eggs, which remain in the brood pouch until the embryonic development is
terminated. After the young daphnids are released they start on their own life
cycle. Subsequently, the mother may produce additional broods. With increasing age the probability that a daphnid may die increases. Feeding rate,
growth rate, development rate of both juveniles and embryos, as well as the

number of eggs and the probability of death depend on the environmental conditions. The effects of temperature, photoperiod and food on the life history
of a daphnid have been introduced into the simulation model by several equations which have been fitted from the results of the breeding experiments
using single daphnids. For the feeding rate which depends on the temperature,
the food concentration and the length of the animals, we applied the equations
published by Geller (1975). In order to take into account the variability of
the life history parameters, each daphnid obtained another individual growth
rate, development rate, etc. by means of a random generator which drew the
individual value for each property from the corresponding probability distribution. The latter had been derived from the empirically observed distribution.

3. Simulation

The life cycle of an animal has been written in the algorithmic language
SIMULA. A flow chart of the program is illustrated in Fig. 1. The life of
a single model daphnid proceeds in regular time steps. At the beginning of
each time step the actual values of the life history parameters are calculated and then fixed for the subsequent time interval; this is repeated until the daphnid dies. The development of juveniles and embryos is a continuous process, which in this program is divided into discrete time units
(= development rate). As shown for insects, the development rate of either
juvenile daphnids or embryos can be summed up after each time step until the
value 1 (= 100%) is reached. Then either the juvenile daphnid becomes mature
or the embryo becomes a juvenile daphnid. After termination of embryonic development, the young daphnids are released and then start on their own life
cycle.

Each daphnid that follows the above life cycle description simultaneously
contribute to the model population. They compete with each other for the
food: the actual concentration of algae, therefore, depends on the population filtering rate (sum of individual filtering rates) and is calculated
every 'day' in the program.

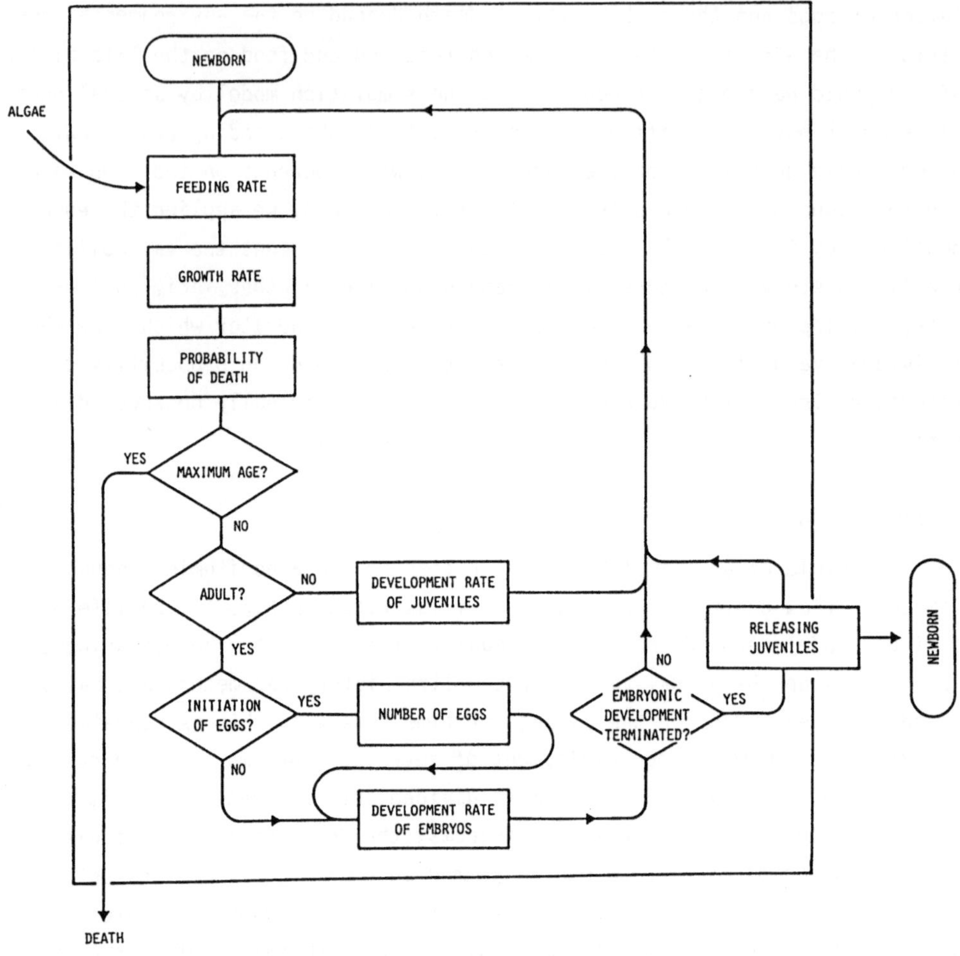

Fig. 1: Flow chart discribing the life cycle of an individual daphnid

4. Results

Fig. 2a shows the result of a simulation under constant conditions (25°C, LD 14:10, food = $5 \cdot 10^7$ Chlorella cells/d). Time units are days, corresponding to the time, at which laboratory measurements were made. The population started with five animals, increased up to a maximum number of daphnids and then decreased to oscillate round an equilibrium density. This result clearly demonstrates that the population dynamics can be simulated without using any of the usual population parameters (e.g. carrying capacity, intrinsic rate of natural increase). Population growth rate and equilibrium density are solely results of all individuals, each 'living' its own life, but competing for the same food source.

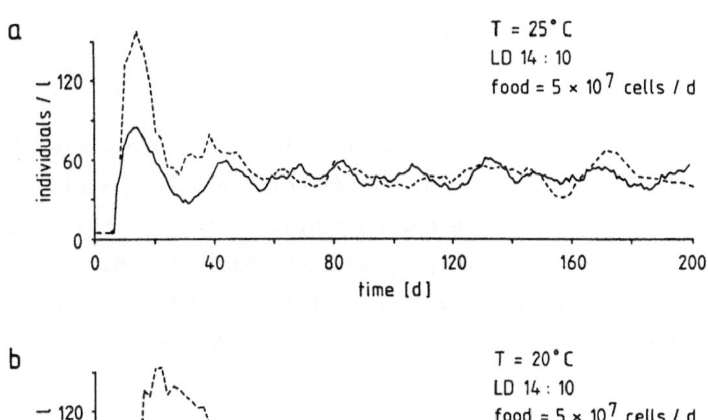

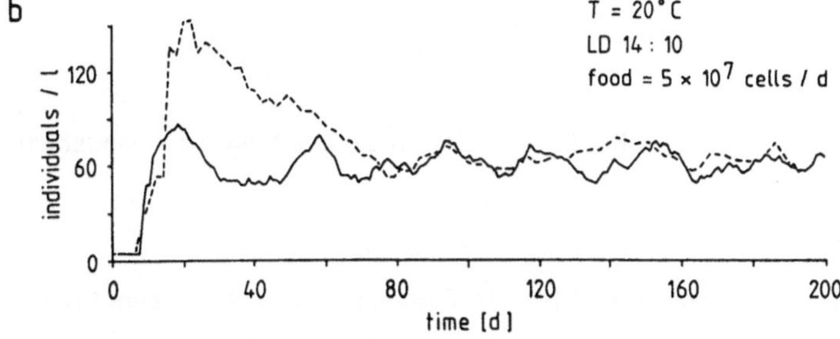

Fig. 2: Broken lines: results of population experiments in the laboratory:
the daphnids were kept in 1l medium, which was changed every second
day. Twice a week the daphnids were counted and fresh algae were
added every day.
Solid lines: simulation results of the population dynamics of
Daphnia magna, whereby the environmental conditions were chosen,
corresponding to those of the experiments.
Fig. 2a: Experiments and simulations at 25°C
Fig. 2b: Experiments and simulations at 20°C

A comparison of the simulated population with the experimental population
(Fig. 2a) reveals that soon after the beginning of the experiment the maximum density is somewhat underestimated by the simulation, but the equilibrium density agrees closely. Similar results are obtained at 20°C (Fig. 2b).
Furthermore, Fig. 2a and 2b show that the influence of temperature on population dynamics may also be predicted accurately. The equilibrium density
at 20°C is higher than at 25°C in both the experiment and the simulation.
The difference between simulation and experiment in the initial phase is
probably caused by the relatively better food conditions: Below the equilibrium density each animal can feed more Chlorella cells per day than in

subsequent periods. Therefore, we suggest that the storage of fat etc. may
be responsible for the higher maximum density in the experiment.

The close agreement between experiment and simulation model clearly indicates
that not only has the life cycle of the daphnid been adequately quantified,
but also that the effects of the environmental factors on the population dynamics may be predicted accurately by this approach. Further this model seems
to be suitable for estimating the impact of anthropogenic factors such as
heavy metals, pesticides etc. on the population dynamics.

Acknowledgements:

We wish to thank Dr. H.T. Ratte for critical reading the manuscript.

References:

Geller, W. (1975): Food ingestion of *Daphnia pulex* as a function of food
    concentration, temperature, animals body-length and hunger. Arch. Hydro-
    biol. Suppl. 48: 47-107

Kaiser, H. (1975): Populationsdynamik und Eigenschaften einzelner Individuen.
    Verh. Ges. Ökol. Erlangen: 25-38

Estimation of Individual Growth Curves from Aggregate Data

Werner Wosniok, Bremen

Zusammenfassung : Zur Bestimmung des individuellen Wachstums
stehen häufig nur Daten über das Wachstum der Population zur
Verfügung. Es wird eine Methode vorgestellt, das Individual-
verhalten aus Populationsdaten zu schätzen. Abhängig vom Problem wird dazu das Wachstum durch einen Markoff- oder Semi-
Markoff-Prozess modelliert. Das Vorgehen wird durch ein Beispiel illustriert.

Summary : For the analysis of individual growth frequently
only data about the growth of the population is available. A
method is presented which allows the estimation of individual
behaviour from population data. For that purpose growth is
modelled by a Markov or a semi-Markov process, depending on
the particular problem. The procedure is illustrated by an
example.

1.   Introduction

The growth of individuals is, besides being an important characteristic of the species, one of the factors which determine
the dynamics of a biological population. In order to design a
population dynamics model or a model for the study of control
mechanisms some information about individual behaviour is necessary. In many cases, however, especially when dealing with
wild-living populations, individual growth histories are not
available, but the only data that can be collected is of an
aggregate type giving frequencies of growth states in samples
from the population taken at several consecutive time points.

This paper introduces a method which allows the estimation of
individual growth curves from aggregate data under some general
assumptions. The term 'growth' refers to changes with time of
a continuous (length, weight) or a discrete variable (stage of
development). The incorporation of covariables is possible so

that relations between e.g. living condition (described by co-variables) and growth can be analysed.

Methods for the analysis of aggregate data with the aim to find the law of individual behaviour have been proposed repeatedly. A recent paper is that by KALBFLEISCH, LAWLESS, and VOLLMER (1983). KALBFLEISCH and LAWLESS (1985) discuss the analysis of panel data by methods which can be applied to aggregate data in a similar way. Both papers contain further useful references. While most of the papers cited assume a Markov process controlling the growth process, the Markov assumption is not essential for the method presented in this paper.

2. Structure of Data

Even with growth measured by a continuous variable, any single value often is recorded only as lying in an interval $[l_{i-1}, l_i)$, $i=1,2,\ldots,I$. Therefore, measurements of either a continuous or a discrete growth variable L will be treated uniformly as resulting in counts $x_i(t_k)$, which denote the number of individuals with growth variable values in $[l_{i-1}, l_i)$ at time t_k, $k=1,2,\ldots,K$. The class $[l_{i-1}, l_i)$ is referred to as class i. For a discrete variable L, class i refers to the i-th value of L.

The sampling scheme from which the counts x_i emerge determines one part of the likelihood function used in the estimation procedure. Various types of samples can be found ; e.g.

- small samples from an 'infinite' population,
- large samples from an 'infinite' population,
- complete observations of a closed finite population.

The distinction between large and small samples is somewhat arbitrary. Large samples produce good estimates for the class occupation probabilities at each time t_k, whereas small samples do not. For brevity further developments will be given only for small samples from an 'infinite' population.

Sampling is assumed to not influence the growth of the population. Fluctuations in the sample sizes from time to time might

be a result of immigration into or emigration out of the observation area, or of birth and death. These changes could be accounted for by additional model terms (which require knowledge
or assumptions about corresponding mechanism).

3. Modelling Growth

The method discussed here bases on the assumptions that

- all individuals behave statistically independent from one
  another,
- migration and death occur with identical probabilities in
  all classes,
- there are no birth events during the sampling period.

In order to make inferences about individuals possible, the
population growth must be modelled in terms of individual growth.
Growth of an individual means that it moves from class i to another class j. This moving is governed by probabilities

$$p_{ij}(t) = \text{Prob }\{\text{individual in j at t} \mid \text{in i at 0}\},$$
$$P(t) = (p_{ij}(t)), \quad i,j = 1,2,\ldots,I ;$$

while the probability to occupy class i at t is

$$p_i(t) = \text{Prob }\{\text{individual is in i at t}\}$$
$$p(t) = (p_i(t)), \quad i = 1,2,\ldots,I .$$

If P is chosen to be

$$P(t) = \exp(Qt) = \sum_{n}^{\infty} (Qt)^n/n!$$

with Q an IxI-matrix with infinitesimal transition rates q_{ij}
as entries ($q_{ij} \geq 0$ for $i \neq j$, $q_{ii} = \sum_{j \neq i} q_{ij}$, $q_{ij} = 0$ for impossible
transitions), and an initial distribution p_0 is given, the
growth process is modelled by a time-homogeneous Markov process.
The time t_0, for which p_0 is valid, may or may not coincide
with the first observation time t_1. For $t_0 \leq t$ we have

$$p(t) = p_0 \, P(t-t_0), \tag{1}$$

and the expected individual growth curve is obtained from

$$E(L(t)) = \sum_i 0.5 \; (1_{i-1} + 1_i) \; p_i(t) \; , \qquad\qquad (2)$$

where

$$p(t) = e_1 \; P(t),$$

e_1 denoting the first 1xI unit vector.

Assuming a homogeneous Markov process as model for the growth
process is equivalent to assuming that the waiting time in each
class is exponentially distributed. This implies that, no matter
what time has been spent in class i, the expected time until
exit from i is the same. Further, exponential waiting times are
inconsistent against collapsing classes : the time to pass several consecutive classes is no more exponentially distributed.
For some problems, these properties might be unsatisfactory.
If so, a more appropriate model can be defined by specifying
adequate (non-exponential) waiting time distributions F_{ij} for
each transition. The growth model then is modelled by a semi-
Markov process. Here special considerations become necessary
to account for the time spent by the individuals in their initial
states before t_0. It will be assumed, however, that the history
before t_0 can be accounted for by additional model terms, and
consequently eq. (1) and (2) will be used furtheron.

As an example consider individuals which grow by moving from
class 1 to I successively without skipping a class. Here only
$F_{i,i+1}$, $i=1,2,\ldots,I-1$ must be specified ($F_{Ij} \equiv 0$). Let $F_i := F_{i,i+1}$.
Then the probabilities p_{ij} for $i \leq j$ are obtained as differences
of convolutions of the F_i :

$$p_{ij}(t) = \begin{cases} 1 - F_i(t) & i = j \\[1ex] F_i * F_{i+1} * \cdots * F_{i+j-1}(t) - \\ F_i * F_{i+1} * \ldots * F_{i+j}(t) & i < j < I \\[1ex] F_i * F_{i+1} * \cdots * F_{i+j-1}(t) & i < j = I \\[1ex] 0 & i > j \end{cases}$$

Here '*' denotes convolution :

$$F_i * F_j(t) = \int_0^t F_i(t-s) \, dF_j(s) \; .$$

Transition probabilities for other growth structures are derived in a similar way.

4. Estimation

The probabilities p_{ij} and p_o are unknown and must be estimated from data. If in a particular situation p_{ij} and p_o are to depend on covariables z_m, the parameters θ_m relating p_{ij} and p_o with z_m must be estimated instead. Typically a linear $(u=\Sigma \; \theta_m \cdot z_m)$ or a loglinear $(u=\exp(\Sigma \; \theta_m \cdot \dot{z}_m))$ predictor u is used to model the entries of Q or the unknown parts of F_{ij}, respectively. The estimated value for the unknown parameter θ is the value $\hat{\theta}$ for which the log-likelihood function

$$\ell = \sum_k \sum_i \; x_i(t_k) \; \ln(p_i(t_k))$$

attains its maximum. Maximisation of ℓ can be performed e.g. by the scoring method (cf. RAO (1973), p. 37o). The scoring method requires first derivatives of p_{ij} and p_o with respect to θ. In the Markov model the derivatives of p_{ij} can be computed using an approach prposed by KALBFLEISCH, LAWLESS, and VOLLMER (1983), given that Q has distinct eigenvalues.

5. Illustration

Growth data from the antarctic krill Euphausia superba Dana (CLARKE and MORRIS (1983), RICKETTS, British Antarctic Survey, Cambridge/UK, pers. comm.) was analysed using the method presen-ted. Samples were taken for six days in 4-h intervals. 30 samples with sizes between 100 and 400 individuals were used for the analysis. The growth variable was individual length in mm, cate-gorized in 20 classes with midpoints between 21 and 52 mm. The original data was corrected for observer-specific measurement deviations (WATKINS, MORRIS and RICKETTS (1985), RICKETTS, pers. comm.).

The initial distribution p_0 was modelled by a logistic distribution

$$p_i(t_0) = 1/(1+\exp(-\theta_{p1}-\theta_{p2}\cdot l_i)) - 1/(1+\exp(-\theta_{p1}-\theta_{p2}\cdot l_{i-1})).$$

As a first approximation all individuals were assumed to have entered their initial state just at t_0. Waiting times were modelled by gamma distributions

$$F_i(t) = F(t;\theta_0,\theta_i) = \theta_0^{\theta_i} \int_0^t e^{-\theta_0 x} x^{\theta_i-1}\, dx / \Gamma(\theta_i) .$$

The use of exponential distributions for F_i was found to be not adequate because of the non-memory property of exponentials and the fact that, when using exponentials, fundamental parts of the growth process would be modelled depending on the arbitrarily chosen class limits. Gamma distributions do not establish such a dependency as they are consistent against collapsing classes if the scale parameters θ_0 are identical :

$$F(t;\theta_0,\theta_i) * F(t;\theta_0,\theta_j) = F(t;\theta_0,\theta_i+\theta_j) \quad .$$

A sketch of the results is given by figures 1 and 2. Method and result will be discussed in more detail in a forthcoming paper.

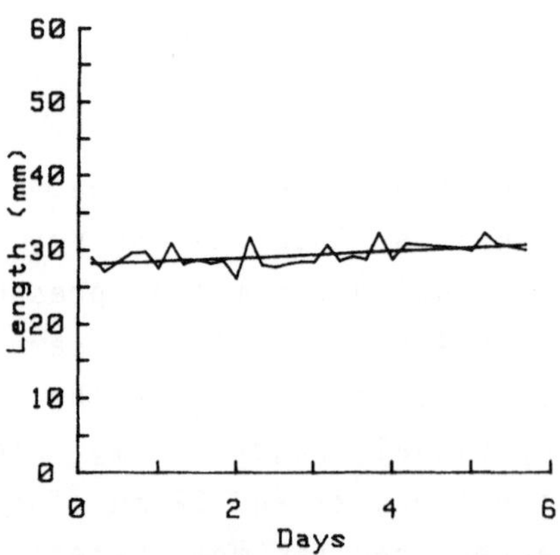

Fig. 1 : Observed vs.
predicted population growth

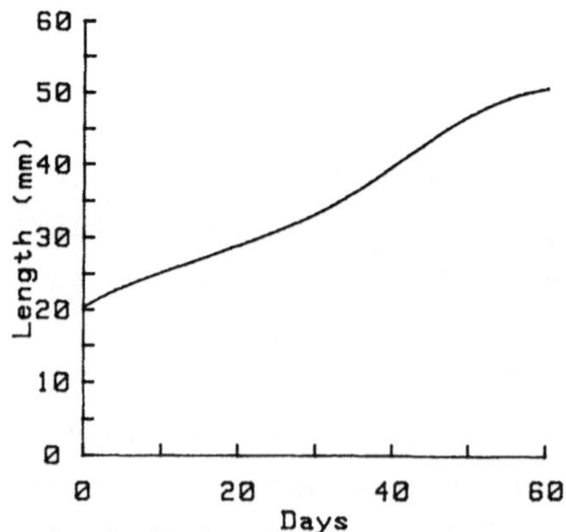

Fig. 2 : Predicted individual
growth starting in class 1
extrapolated to 60 days

References

CLARKE,A., MORRIS,D.J. (1983) : Towards an energy budget for
   krill : the physiology and biochemistry of Euphausia
   superba Dana. Polar Biol. $\underline{2}$, 69 - 86

KALBFLEISCH,J.D., LAWLESS,J.F. (1985) : The analysis of panel
   data under a Markov assumption. Journal of the American
   Statistical Association $\underline{80}$, no. 392, 863 - 871

KALBFLEISCH,J.D., LAWLESS,J.F., VOLLMER,W.M. (1983) :
   Estimation in Markov models from aggregate data.
   Biometrics $\underline{39}$, 907 - 919

RAO,C.R. (1973) : Linear statistical inference and its
   applications. 2nd edition. John Wiley and sons, New York

WATKINS,J.L., MORRIS,D.J., RICKETTS,C. (1985) : Nocturnal
   changes in the mean length of a euphausiid population :
   vertical migration, net avoidance, or experimental
   error ? Marine Biology $\underline{75}$

Sleeping Stem Cells – A Model of Of Stem Cells under Continuous Stress

Günther Pabst, Ulm

Summary. A model of stem cell proliferation and regulation was developed and simulated in which cells in G_0 after mitosis only gradually become suscepti-ble to be stimulated into a new cell cycle, the length of this resting phase of limited stimulability being dependend on the stem cell's life history. If in addition cells are assumed to lose the self-reproducing capability and thus their stem cell characteristics if this resting phase has been short-ened too frequently by early stimulation, the model reproduces the sudden breakdown of the hematopoietic system sometimes observed in animals during a continuous irradiation at a time long after development of a new steady state.

Zusammenfassung. Ein Regulationsmodell der Stammzellproliferation wurde entwickelt und simuliert, bei dem Zellen in G_0 nach der Mitose nur langsam wieder für eine Stimulation in einen neuen Zellzyklus empfänglich werden, wobei zudem die Länge dieser Ruhephase verminderter Stimulierbarkeit von dem vorausgegangenen Schicksal der Stammzelle abhängt. Unter der zusätzlichen Annahme, daß Zellen ihre Selbstreproduktionsfähigkeit und damit ihre Stamm-zelleigenschaft verlieren, wenn diese Ruhephase zu häufig aufgrund einer frühen Stimulation verkürzt worden ist, ist das Modell in der Lage, den plötzlichen Zusammenbruch des haematopoetischen Systems nachzuvollziehen, wie er in Tieren bisweilen eine lange Zeit nach Beginn einer chronischen Bestrahlung und Ausprägung eines neuen Gleichgewichtszustands beobachtet werden konnte.

Introduction: Proliferation of stem cells

Cell systems with continual losses of cells, an example of which is the hematopoietic system, need a constant influx of new cells to keep up their numbers. These are supplied by dividing stem cells, cells with self-replica-ting ability.

For the sequel we will restrict our considerations to stem cells of the hematopoietic system. During steady state half of their daughter cells after division will retain stem cell properties and the other 50% will give rise to progeny of the erythropoietic, the granulopoietic, the thrombopoietic, or possibly also the lymphopoietic system.

According to the need for functional cells the proportion of stem cells differentiating is regulated as well as the proliferation rate of the stem

cells - and there is ample room for increased proliferation of stem cells,
because under normal circumstances only about 10% of all stem cells prolife-
rate at any one time.

All the other cells are considered to be in a so-
called G_0-phase (1), situated somewhere between a
postmitotic rest phase (dichophase) and a presynthetic phase, see Fig. 1. However, this G_0-phase
is just an operational term: it has not been
possible to attribute biochemical actions to the
G_0-phase or to elucidate its physiological role
besides functioning as a kind of reservoir. We
will therefore consider G_0 not to be a part of
the cell cycle.

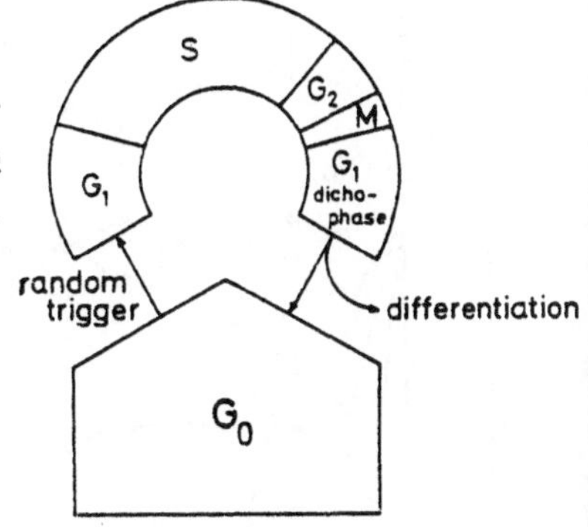

Fig. 1. Cell cycle and G_0-phase

First stage of a model of resting and proliferating stem cells

We assume that differentiation of stem cells occurs after completion of a
postmitotic resting phase (dichophase). All cells that did not differentiate
and thus retained stem cell properties will then enter G_0. If need arises
they are triggered randomly from the G_0-phase into a new reproductive cycle
starting with a presynthetic phase, compare Fig. 1.

Under continual stress, e.g. during continuous irradiation, the triggering
force will increase thus stimulating more stem cells into proliferation.
Such a model corresponds to the current understanding of the stem cell
system's regulating mechanisms.

Results on continuous irradiation

In mice the colony forming units in spleen (CFU-S) have characteristics of
stem cells. If now mice are continuously irradiated at low dose rates, the
number of CFU-S will decrease, but - if the rate of daily irradiation has
been low enough - a new steady state will be achieved, see Fig. 2 (2). Such
a steady state during a situation of continual stress is only possible with
an increased proliferation of stem cells in order to compensate for the
losses.

Similar results were observed after irradiation of dogs (3,4). Of 24 dogs
irradiated continuously with 5 R/day two died within 484 days of anemia
after a progressive decline in all cellular blood elements. These animals
obviously did not succeed in establishing a new steady state. In the other
22 dogs blood cell counts stabilized after an initial decrease. 11 dogs died

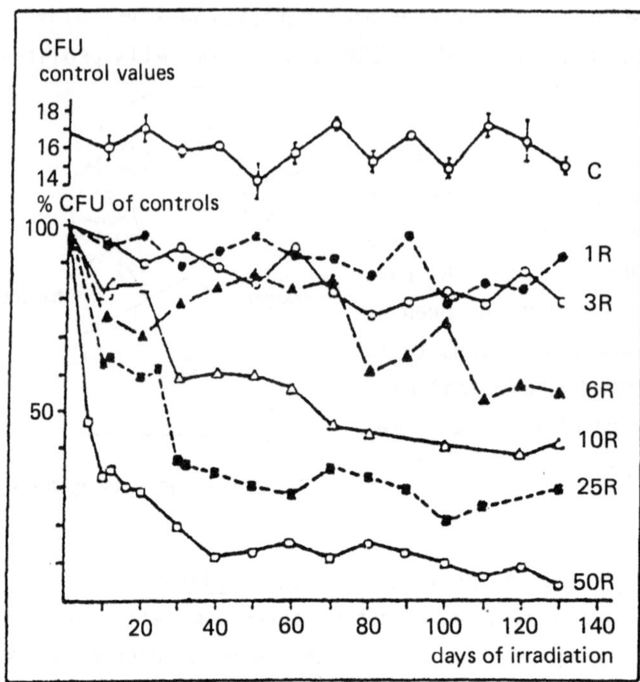

Fig. 2. Per cent CFU values in mice continuously
irradiated at dose rates from 1R—50R/day.
Kalina et al. Folia Biol (Praha) **21** (1975), 165

after total doses of between 10380 R and 15460 R with carcinomas or of
causes that could not be related to the hematopoietic system. The other
11 dogs that died after intermediate total doses of 4945 R to 9745 R died
because of myeloproliferative disorders (MPD).

In some of these dogs shortly before death a steep increase of the numbers
of white blood cells and/or thrombocytes could be noted. This effect could
result if a cancerogenous clone of cells had developed that did not respond
to any regulatory feed-backs anymore.

In some other dogs that died of MPD death came with a sudden fall in all
blood cell numbers as if a limited supply of (stem) cells had run out,
compare Fig. 3. It was the purpose of the research reported here to try out
another explanation for this observation.

Second stage of a model for the regulation of stem cell proliferation

In a refinement of the model we assumed that after a dichophase of fixed
length the stem cells only gradually become more and more susceptible to be
triggered into a new cell cycle. After a certain time span t_N after mitosis

has elapsed, t_N differing between individual cells, all cells are assumed to
be equally susceptible for the random trigger, compare Fig. 4.
Furthermore, the length of time t_N' that cells are not fully available for
stimulation into a new reproductive cycle was also assumed to depend on the

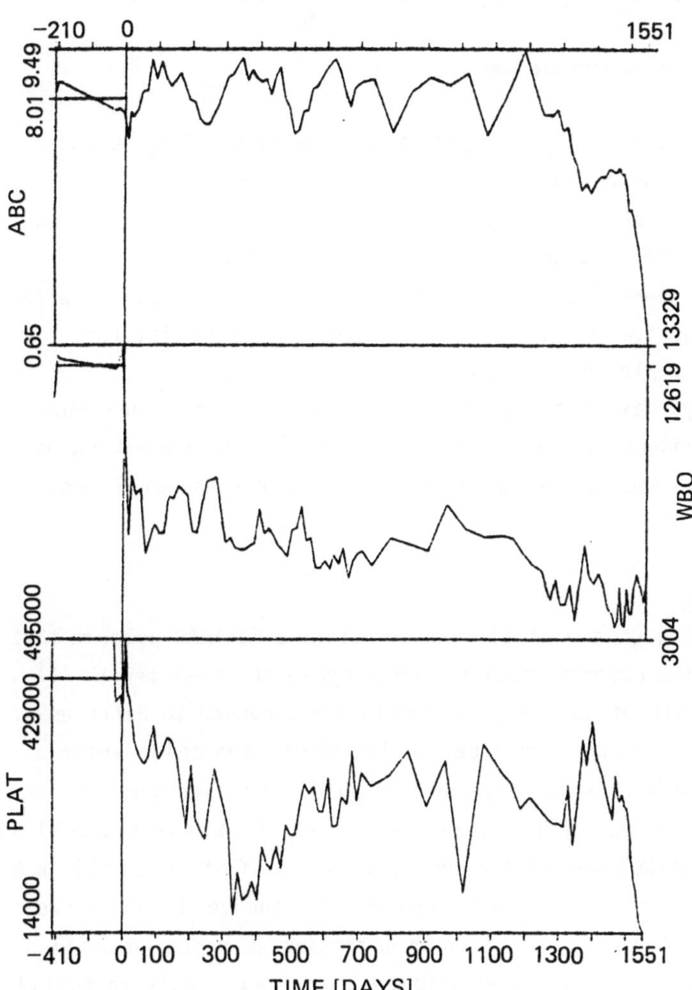

Fig. 3. Blood cell counts of a dog during
continuous ^{60}Co γ-irradiation.   Data
courtesy of T.E.Fritz, Div. Biol. Med.
Res., Argonne Natl. Lab., Argonne, Ill.

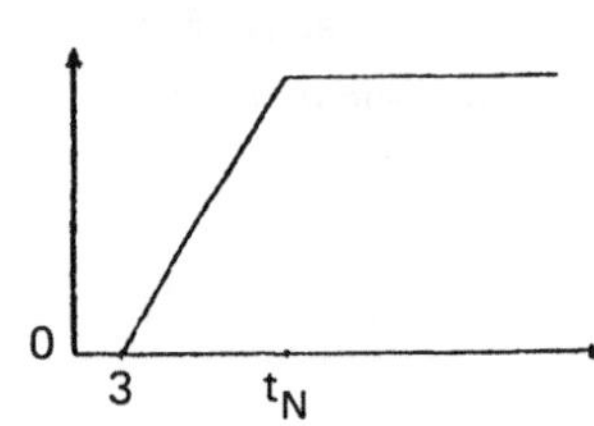

Fig. 4. Probability for stimulation into a new cell cycle.

$$\text{①} \quad t_N' := \max\left(\left(1-\frac{t_{stim}-t_N}{t_N-2}\right)t_N,\ 1.15\,t_N-0.01\,(t_{stim}-t_N),\ 3.0\right)$$

$$\text{②} \quad t_N' := \max\left(1.15\,t_N\exp(-0.004\,(t_{stim}-t_N)),\ 3.0\right)$$

$$\text{③} \quad t_N' := \frac{3\,(t_{stim}-t_N)+1.15\,(60+t_N)\,t_N}{60+t_{stim}}$$

$$\text{④} \quad t_N' := \max\left(\frac{1}{9}\left(11.5\,t_N-3+\left(\frac{3}{t_N}-1.15\right)t_{stim}\right),\ 3.0\right)$$

Fig. 5. Formulae used.

time of stimulation t_{stim} into the last cycle and the value of t_N during the last G_0-phase. It was assumed that the t_N in the new G_0-phase (t_N') will be greater, equal or less than t_N if triggering occured earlier than t_N, i.e. for $t_{stim}<t_N$, at t_N, or later than t_N ($t_{stim}>t_N$), respectively.

Note that t_N and t_{stim} count from the time of mitosis. The time period after dichophase and up to t_N, i.e. the period of limited susceptibility for stimulation into a new cycle, shall be called N-phase.

The N-phase in some ways may be compared to the sleep in animals and man. After a days work (reproductive cycle) one goes to sleep. Before waking up by yourself (at time t_N) you may be awakened for new work - and will need more sleep the next night.

Simulation of the model

The model was represented on a digital computer using a FORTRAN-program that simulated 1400 cells. The program assumes a cell cycle of fixed length (11 hours). After mitosis half of all daughter cells are supposed to differentiate and eventually become functional cells. The other stem cells enter a dichophase (3 hours) and afterwards N-phase. At each simulation time point cells are chosen at random out of all stem cells in G_0. If a selected cell is not in N this cell is stimulated into a new cell cycle. If it is a cell in N an additional random number is drawn and compared with the cell's relative stimulability (0 at 3h, 1 at t_N). This random selection is repeated until a sufficient number of stem cells has been triggered into cell cycle to fulfill the needs for functional cells - assuming that half of the daughter cells after mitosis differentiate. For cells stimulated into the reproductive cycle the new length t_N' of the next N-phase is calculated immediately.

Four different versions of the model were constructed according to the formu-

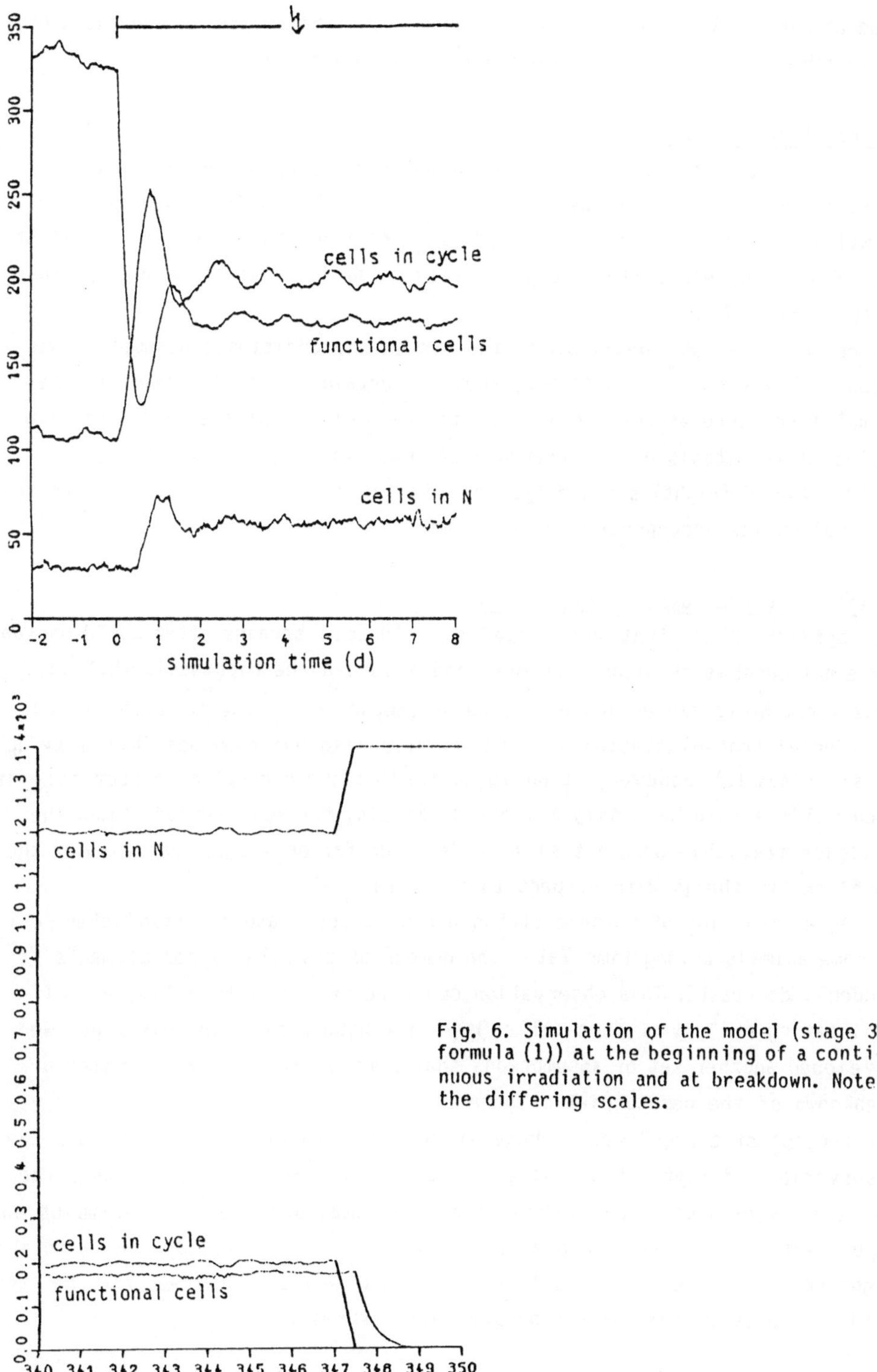

Fig. 6. Simulation of the model (stage 3, formula (1)) at the beginning of a continuous irradiation and at breakdown. Note the differing scales.

lae of Fig. 5. All versions however produced similar results and therefore only simulation runs using formula ① are depicted here.

Third stage of the model

Under regular conditions this kind of model works well. During continuous stress a new steady state develops, see Fig. 6 (left). The cells however continuously get too little "sleep" ($t_{stim} < t_N$) and therefore the number of cells in N increases almost exponentially - faster with formula (1), slower with formula (4).

To reproduce a fast breakdown of the system one additional assumption was needed: All stem cells with a t_N above a certain limit, 210 hours in the simulation, were assumed to lose their stem cell properties and both daughter cells after mitosis will differentiate thus reducing the number of stem cells. Fig. 6 (right) shows a typical simulated curve of breakdown when this assumption was incorporated into the model.

Relation to the Hayflick hypothesis

L. Hayflick showed that human fetal diploid cell strains after 50 ± 10 passages were not capable of mitosis anymore and advanced the hypothesis that this fact might be an expression of ageing or senescence at the cellular level (5). Serial transplantation of CFU-S in mice also has been possible only up to six times (6). However, it would contradict the concept of a stem cell if such cells should have only a limited capacity for self-reproduction. The evidence available does not allow a decision for or against validity of the Hayflick hypothesis with respect to stem cells (7).

During a situation of chronic stress a new steady state is established, but in some animals a long time later the number of cellular blood elements suddenly decreases. This observation could be explained by a limited self-replicating ability of stem cells (Hayflick hypothesis). In this paper we developed another set of assumptions that similarly would provoke such a breakdown of the hematopoietic system.

Our concept of a regulated N-phase on the other hand could also explain the observations of Hayflick and others, because in their in vitro systems the cells are kept under a continuous, almost maximal proliferative stimulus. An experiment with serial transplantation of CFU-S in mice would be able to shed light on our hypothesis, if a longer time span would be allowed for the stem cells to recover from their transplantation stress.

Acknowledgements
The research reported here was performed at the University of Ulm, Fed. Rep.
Germany, in the Department for Clinical Physiology and Occupational Medicine
(Director Prof. T.M. Fliedner) and was supported by the Deutsche Forschungs-
gemeinschaft through Sonderforschungsbereich 112.

References
1. L.G.Lajtha: On the concept of the cell cycle. J Cellul. Comparat. Physiol.
   62, Suppl.1 (1963), 143-145 (discussion pp.151ff)
2. I.Kalina, M.Praslička, L.Marko, V.Krasnovská: Effect of continuous irradiation upon bone marrow haemopoietic stem cells in mice. Folia biol.
   (Praha) 21(1975), 165-170
3. T.E.Fritz, W.P.Norris, D.V.Tolle: Myelogenous leukemia and related myeloproliferative disorders in beagles continuously exposed to ^{60}Co γ-irradia-
   tion. Unifying concepts of leukemia, R.M.Dutcher, L.Chieco-Bianchi, eds.,
   Karger, Basel 1973, pp.170-188
4. W.P.Norris, S.A.Tyler, G.A.Sacher: An interspecies comparison of responses
   of mice and dogs to continuous ^{60}Co γ-irradiation. Biological and environmental effects of low level radiation, Proc. Symp., Int. Atomic Agency,
   Vienna 1976, vol.1, pp.147-156
5. L.Hayflick: The limited in vitro lifetime of human diploid cell strains.
   Exp. Cell Res. **37**(1965), 614-636
6. H.S.Micklem, D.A.Ogden: Ageing of haematopoietic stem cell populations in
   the mouse, Stem cells of renewing cell populations, A.B.Cairnie, P.K.Lala,
   D.G.Osmond, eds., Academic Press, New York 1976, pp.331-341
7. U.Reincke, H.Burlington, E.P.Cronkite, J.Laissue: Hayflick's hypothesis:
   an approach to in vivo testing. Fed. Proc. **34**(1975), 71-75

A Mathematical Method of Modelling and Simulating Biological Structure Control Systems

Thomas Vogelsaenger , Siegen

Zusammenfassung: In dieser Arbeit wird eine Methode vorgestellt, mit deren Hilfe es möglich ist, unter Anwendung formaler Mathematik dynamische Strukturen biologischer Systeme zu modellieren und simulieren. Diese Systeme sollen "Strukturregelkreise" genannt werden, weil die entsprechenden Modelle auf der Theorie beruhen, daß alle dynamischen biologischen Strukturen das Ergebnis von biologischen geschlossenen Strukturregelkreisen sind, die sowohl stabil als auch instabil sein können. Die große Komplexität biologischer Systeme erfordert die Anwendung unterschiedlichster mathematischer Methoden, was jedoch keine Einschränkung für die vorgestellte Methode bedeutet, die ursprünglich für die Modellbildung und Simulation von Tumorwachstum und –behandlung angewendet wurde mit dem Ziel, optimale Tumortherapien durch Computersimulation zu ermitteln.

Summary: In this paper a method will be introduced that has the capability to modell and simulate the dynamic structures of biological systems with the help of formal mathematics. These systems will be named " structure control systems" because the corresponding models are based on the theory, that all dynamic biological structures are the result of biological structure closed loop circuits which are either stable or unstable. The great complexity of biological systems requires the application of many different mathematical methods, which is no restriction for the presented method that was originally applied to the modelling and simulation of tumor growth and –treatment with the aim to determine the optimal tumor therapy by computer simulation.

1. Introduction

The mathematical modelling and simulation of tumor growth and treatment /1-3/ requires the simultaneous application of different mathematical methods as formal language theory, automata theory, (deterministic and stochastic) systems theory, field theory, information theory and control theory. In this paper the control theory is the main point of view which will be motivated in the following.

Control Mechanisms of structure building processes for example the development of road networks or power supplying networks and the development of
biological systems may be termed as " structure building processes". System
variables affecting the structure building process as feedback signals,form
closed loop circuits in the terms of control theory and may be defined as
"structure control systems". The analysis of biological "structure control
systems" is of great importance because it seems to be obvious, that the
dominant structure influencing system variables are also of substantial importance for the cource of diseases.

In /1-3/ it has been shown,that the 3-dimensional modelling of tumor growth
enables not only the simulation of "in vitro"- and of "in vivo"-experiments
/4-7/. Furthermore it was possible to schedule a combined tumor therapy for
fictitious tissues. The aim of this work is to give an insight into the
special mathematical method of modelling biological processes on a cellular
level. Thus, in the future it may be possible to build modells of different
tissues as skin, liver or brain.

2. Modelling of Structure Control Systems

In this section the modell of an overall "structure control circuit" for
cell multiplying processes will be derived, which is defined as an extended
approach of the systems theory /8-11/.

The set of all biological cells ZG may be devided into the subsets ZG_i
whitch form a volume Ω_i in the 3-dimensional space R^3 with the space structure $S(\Omega_i, ZG_i)$. From the biological point of view the set of all structure
elements ZG_i has to be subdivided into strukture components ZG_i^j representing the different characters of bone-, tissue- or blood-cells.

$$ZG \quad = \{ \ ZG \quad : ZG \quad \text{is a biological cell} \ \} \tag{1}$$

$$ZG_i = \{ \ ZG_i \ : ZG_i \ \text{is a cell in the Volume } \Omega_i \ \} \tag{2}$$

$$S \quad = \{ \ S \quad : S \ = \ f \ (\ \Omega_i, \ ZG_i) \ \} \qquad . \tag{3}$$

The dynamic variations of biological structures are of special interest.

Therefore one may write the space structure in the form

$$S = f\,(\Omega_i,\, \mathbf{ZG}_i,\, t) \qquad \text{or} \qquad S = f\,(\Omega_i,\, ZG_i^{\,j},\, t) \quad . \tag{4}$$

In this case it is necessary to describe all ZG_i by a set $\boldsymbol{\theta}$ of n variables $\theta_i^{\,j}$ which may be represented by a vector $\underline{\theta}$

$$\theta_i = \bigcup_{j=1}^{n} \theta_i \qquad \text{and} \qquad \underline{\theta} = (\,\theta_i^1,\, \theta_i^2,\, \ldots\,,\, \theta_i^n\,)^{\,T} \tag{5}$$

$$ZG_i \quad \Longleftrightarrow \quad \underline{\theta}_i(ZG_i) \tag{6}$$

A cell multiplying process PZ_i may be defined in the volume Ω_i by the cartesian product of the structure S and the time T from which the following structure S' is generated

$$PZ_i\,:\, S\,(\Omega_i,\, \mathbf{ZG}_i\,)\, X\, T\, \to\, S'\,(\Omega_i\,,\, \mathbf{ZG}_i\,) \tag{7}$$

where T means the positive time axis. Dynamic structure processes may also be derived by the sum of local relations pz_i in the volume Ω_i, applied to every element of the cell set

$$pz_i\,:\, \underline{\theta}\,(ZG_i)\, X\, T\, \to\, \underline{\theta}'\,(ZG_i) \tag{8}$$

and with the relation

$$\mathbf{ZG}_i \quad \Longleftrightarrow \quad S\,(\,\Omega_i,\, \mathbf{ZG}_i\,) \tag{9}$$

follows

$$PZ_i\, =\, pz_i\,(ZG_i)\,:\, \mathbf{ZG}_i\, X\, T\, \to\, \mathbf{ZG}_i \tag{10}$$

This means, that cell multiplying processes may be described by structure processes which are derived from global transformations (equ. 7) as well as by local transformations of each individual element ZG_i in the volume Ω_i. Some of these cell multiplying processes are for instance ontogenetic processes or processes which are related to wound healing with the formation of new blood vessels or cell multiplying processes of malignant or non malignant tumors.

Furthermore it is possible to describe cell multiplying processes by vectors with PZ_i defined as a matrix and IN representing the set of natural numbers:

$$PZ_i \quad : \quad \mathop{X}_{pi=1}^{2} \; IN_{pi} \quad \rightarrow \quad A \tag{11}$$

A is representing a set of transition conditions. Equation (7) may also be written as

$$\underline{\theta}_i(t') \; = \; PZ_i \; * \; \theta_i(t) \quad , \quad t' > t \quad \text{and} \quad t,t' \in T \; . \tag{12}$$

In this formulation equation (7) is reduced to a multiplying operation. A restriction of (equ. 12) is, that (in opposite to (equ. 8)) it is not possible to include the different mathematical methods as mentioned in section 1.

A "structure control circuit" is characterized in the way, that a cell multiplying process reaches a genetic given structure in a restricted time (Fig. 1). This structure may be defined as reference input of the control circuit.

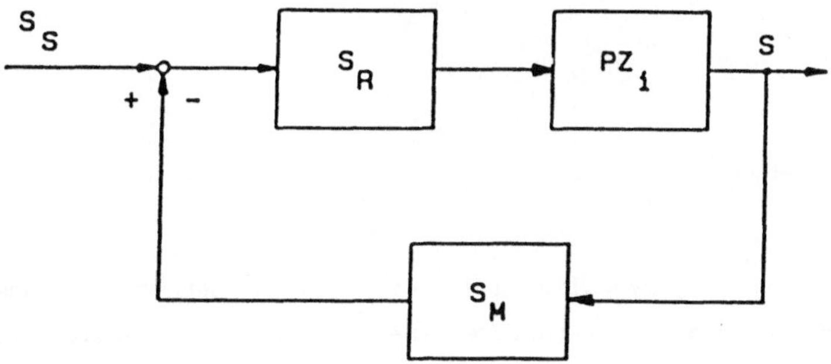

Fig. 1: Diagram of a "structure control circuit".
S_S : reference input structure; S : actual structure; S_R: structure controller; PZ_i : cell multiplying process ; S_M : measurement of structure.

If the dynamic structure of a system is known, the equations 7,8 or 12 allow the modelling and simulation of cell multiplying processes. For example the growing structures of cellular spheroids have a strict relation between volume V and the Structure S which may be described by the radius R.

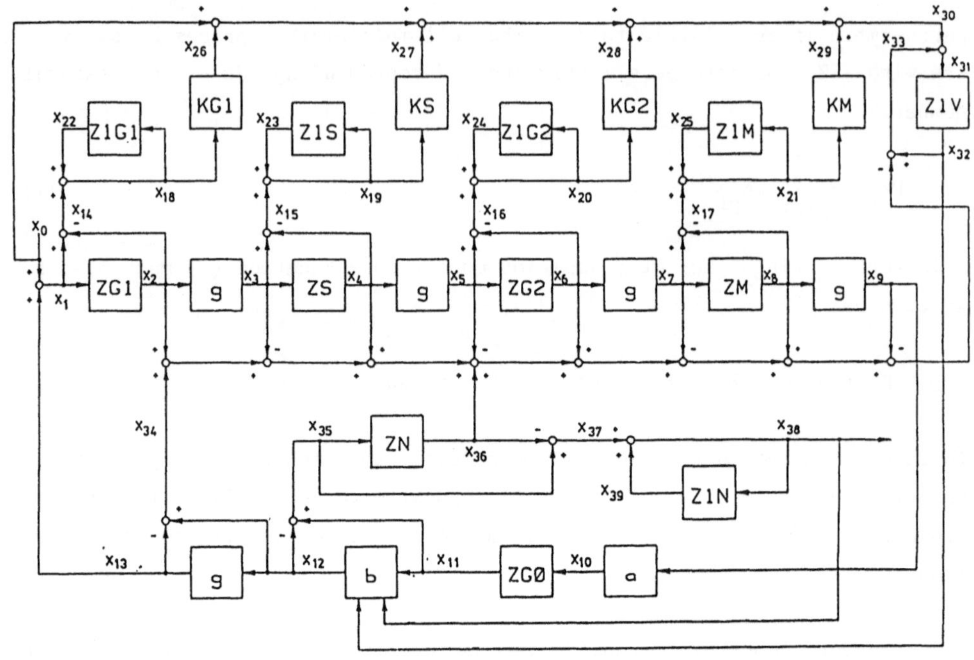

Fig. 2: Diagram of a cell multiplying process in a time discret way

In this case we get according to (equ. 7)

$$S \; -> S' \; , \qquad V = -\frac{4}{3}\,\pi R^3 \qquad \text{and} \qquad (13)$$

$$V = a * e^{(-b * e^{(-ct)})} \; . \qquad (14)$$

Equations (13) and (14) symbolize a direct relation between the time-beha-
vior of the volume and the structure. The same process modelled according
to (equ. 12) leads to a time discrete modell (Fig. 2) where ZG1, ZS, ZG2,
ZM, ZGO and ZN are different cell-cycle phases, g is the cell-loss per time
unit, a represents the doubling process of cells, Z1G1, Z1S, Z1G2 and Z1M
are memory elements which store the actual volume of the cells during the
different cell-cycle-phases and are summed up with the factors KG1, KS, KG2
and KM to the total volume which is accumulated in the memory element Z1V.
The factor b characerizes fraction of proliferating (multiplying) cells
which enter the necroses (cell-death) and are stored in the memory element
ZN. The modelling and simulation according to (equ. 10) are described in
/1-3/. Both, the results corresponding to (equ. 10) as well as to (equ. 12)

are fitting the same time behaviour of the growing tumors as described in
the equations (13) and (14). If it is not possible to derive mathematical
formulations of the structures, which is normal by modelling biological
processes, one has to work with local transformations (equ. 10) on a cel-
ular level. At this level /3/ no restrictions have to be taken into account
for the dynamic and not predictible structures.

3. References

/1/ Düchting, W., Vogelsaenger, Th.: "Aspects of modelling and simulating
     tumor growth and treatment", Journal of Cancer Research and Clinical
     Oncology, No. 105, pp. 1-12, 1983
/2/ Düchting, W., Vogelsaenger, Th.: "Recent progress in modelling and
     simulation of three-dimensional tumor growth and treatment", BioSys-
     tems 18, Elsevier Scientific Publishers, Ireland Ltd, pp. 79-91, 1985
/3/ Vogelsaenger, Th.: "Modellbildung und Simulation von Regelungsmecha-
     nismen wachsender Blutgefäßstrukturen in normalen Geweben und malig-
     nen Tumoren", dissertation at the University of Siegen, in press
/4/ Sutherland, R.M., McCredie, J.A., Inch, W.R.: "Growth of multicell
     spheroids in tissue culture as a model of nodular carcinomas", J.
     Natl. Cancer Inst. 46: pp 113-120, 1971
/5/ Folkman, J., Hochberg, M.: "Self-regulation of growth in three di-
     mension", J. Exp. Med. 138: pp. 745-753, 1973
/6/ Rajewsky, M.F.: "Proliferative properties of malignant cell systems",
     in Altman H. et al. (eds.)"Handbuch der Allgemeinen Pathologie:Tumors
     I,VI/5", Springer Verlag, Berlin, New York, pp. 289-325, 1974
/7/ Carlson, J.: "Tumor models in vitro: A study of proliferation and
     growth in cellular spheroids", Acta Univ. Uppsala 466, 1978
/8/ Burks, A.W.: "Essays on Cellular Automata", University of Illinois
     Press, 1970
/9/ Wunsch, G.: "Zellulare Systeme", Reihe der Wissenschaft, Vieweg,
     Braunschweig, 1977
/10/ Mesarovic, M.D., (ed.): "System Theory and Biology", Springer, Berlin
     Heidelberg, New York, 1980
/11/ Cook, D.J., Bez, H.E.: "Computer Mathematics", Cambridge University
     Press, 1984

Simulation in electrophysiological pharmacology: Specific interactions of antiarrhythmic agents with ion channels of the cardia cell membrane

Dieter Hafner, Florian Berger, Ulrich Borchard; Düsseldorf

Summary. The Beeler-Reuter model (1) for the reconstruction of cardiac action potentials and the Hondeghem-Katzung equations (3) modelling the specific interactions of pharmacological agents with the sodium channel of the cell membrane were combined in order to simulate and analyse experimental results showing frequency- and potential-dependent sodium channel blockade by antiarrhythmic drugs. The method was equally well applied to results obtained from experiments with the calcium channel antagonists diltiazem and nifedipine and the potassium channel inhibitor sotalol.

Zusammenfassung. Durch Kombination von mathematischen Modellen zur Rekonstruktion kardialer Aktionspotentiale (1) und zur Beschreibung der Interaktionen antiarrhythmisch wirksamer Arzneimittel mit Ionenkanalsystemen der Herzzellmembran (3) wurden experimentelle Befunde mit Natriumkanal-Antagonisten durch Simulation analysiert, in denen sich eine frequenz- und potentialabhängige Natriumkanal-Blockade zeigte. Die Methode wurde gleichermaßen zur Beschreibung der Wirkung von Calciumkanal-Antagonisten (Diltiazem, Nifedipin) und Kaliumkanal-Antagonisten (Sotalol) angewendet.

Patients suffering from coronary heart disease are often endangered by cardiac arrhythmias leading to an increase in mortality, which can be reduced by preventive or therapeutic methods. The origin of arrhythmias may be very complex and is often difficult to analyse under clinical conditions. On the other hand there is a large variety of drugs with antiarrhythmic qualities. Therefore clinical use of antiarrhythmic agents is mainly determined by empirical factors, resulting in the consequence that each drug randomly separates a population of patients into groups of responders and nonresponders. In order to supply clinical therapy with rational means, it is a general aim of pharmacological investigations to find the basic mechanisms of drug actions. Considering antiarrhythmic drugs this can efficiently be achieved by using electrophysiological methods. These include registration of intracellular membrane potential, application of voltage clamp and ionselective microelectrodes. Experimental work is supported by mathematical models, which were originally introduced by Hodgkin and Huxley (HH). They were the first to give a formal description of the voltage- and time-dependence of membrane ionic currents underlying the electrical excitation phenomena in nerve cells (2).

In cardiac electrophysiology a number of HH-type models were derived from the vast amount of experimental data accumulated during the past three decades, so that nowadays special models are available for different tissues of the heart: i.e. sino-atrial node, myocardium and Purkinje fibres.

The majority of our investigations were carried out using myocardial papillary muscle of the guinea pig. For a better understanding of our simulations a short review of the Beeler-Reuter (BR) model for the reconstruction of cardiac action potentials is presented here. For details see (1).

The BR-model is relatively simple, as only 4 different ionic currents are incorporated. There are two potassium (K) outward currents (I_{x1}, I_{K1}), a slow inward current (I_{Ca}) carried by calcium (Ca) ions and a fast sodium (Na) inward current (I_{Na}). These currents cross the cell membrane via ionic channels, which are controlled by a voltage- and time-dependent gating mechanism. The model equations are summarized as follows:

$$(1) \quad \dot{V}_m = -\frac{1}{c_m} \sum_{j=1}^{4} I_j \qquad\qquad I_j = G_j \prod_{l=1}^{n_j} s_{j_l}^{k_l} \; f_j(E_j, V_m)$$

$$(2)\text{-}(7) \quad \dot{s}_j = a_j - (a_j + \beta_j)\, s_j \qquad j = 2, 7$$

$$(8) \quad (\dot{Ca})_i = -10^{-7} I_{Ca} + 0.07 \, (10^{-7} - (Ca)_i)$$

Eq.(1) describes the time-dependent changes of the membrane potential, its first time derivative $\dot{V}_m$ being proportional to the sum of the ionic currents I_j (c_m membrane capacity). In general a current is given by the product of a maximal conductivity constant G_j, one or several activation and/or inactivation variables s_{jl}, and a gradient function f_j defining the electrochemical driving force which depends on membrane potential V_m and the ion specific reversal potential E_j. The gating variables s_{jl} range in

the interval (0,1) corresponding to the relative opening of the channel
gate (0: totally closed, 1: totally open). Eq.(2) - (7) define the time-dependent kinetics of the gating variables. The rate constants a_j, β_j are
functions of the membrane potential. There are 6 gating variables altoget-
her: I_{Na}: m,h,j; I_{Ca}: d,f; I_{x1}: x_1; I_{K1} possesses no gating variable. The
reversal potentials E_j of the ionic currents are assumed to be constant.
Only E_{Ca} depends on the dynamics of the intracellular calcium concentration
$(Ca)_i$. The first time derivative of the latter is defined in eq.(8) formulated as an open two-compartment model. On the whole, the model consists of
an ordinary non-linear system of differential equations with 8 state variables. It has to be integrated numerically. Because of its highly different
time constants ranging from 0.05 to 500 ms it belongs to the so called
"stiff" systems. The model gives a good reproduction of the electrical excitation processes of the cell membrane under physiological conditions. It
can be used to describe the pharmacological control experiment, which is
the basis for the evaluation of the drug effects.
For the antiarrhythmic therapy drugs are of interest, which interact specifically with a particular ion channel system, thus reducing a certain ionic
current in a selective and concentration dependent manner. The classical antiarrhythmic drugs are those, which block the fast sodium inward current,
leading to a decrease in depolarisation velocity and impulse conduction velocity. In some cases a prolongation in refractory period can also be observed. For a discussion of these effects in antiarrhythmic therapy see the literature. Ca-antagonists, and, as recently discussed, potassium channel
blocking substances may also efficiently be used for the suppression and
prevention of special types of cardiac arrhythmias.
Using the hypothesis of the modulated receptor a theoretical model was formulated, explaining the way of action of the sodium channel antagonists.
Fig.1 shows a diagram of the model together with the corresponding differential equations according to Hondeghem and Katzung, 1977 (HK): Three different potential dependent channel states are discriminated: the rested (R),
activated (A) or inactivated (I) state. In each state there are different
rate constants for association (k) and dissociation (l) reactions. The occupation of a channel receptor by the drug D leads to channel states R', A'
and I', where the channel is blocked and can no longer participate in conduction of ion molecules.
All the channels run through the different states in a time- and voltage-de-
pendent way as controlled by the BR-model. Experimental findings led to the

assumption that recovery of blocked, inactivated channels (I') is impeded
by a voltage shift of the corresponding h'-kinetics to more negative potentials, thus blocking the "normal" recovery path I'->R'->R.

(9) $\dot{B} = (k_R R + k_A A + k_I I) D - (l_R R' + l_A A' + l_I I')$

(10) $\dot{h} = a_h I + l_R R' + l_A A' - \beta_h (R + A) - (k_R R + k_A A) D$

(11) $\dot{h}' = a_{h'} I' + (k_R R + k_A A) D - \beta_{h'} (R' + A') - (l_R R' + l_A A')$

(12) $\dot{m} = a_m (1 - m) - \beta_m m$

(13) $R = h - A$

(14) $R' = h' - A'$

(15) $A = m^3 h$

(16) $A' = m^3 h'$

(17) $I = 1 - B - h$

(18) $I' = B - h'$

(19) $B = R' + A' + I'$

(20) $1 = R + A + I + B$

(21) $a_m = f_1(V)$

(22) $\beta_m = f_2(V)$

(23) $a_h = f_3(V)$

(24) $\beta_h = f_4(V)$

(25) $a_{h'} = f_3(V + dV)$

(26) $\beta_{h'} = f_4(V + dV)$

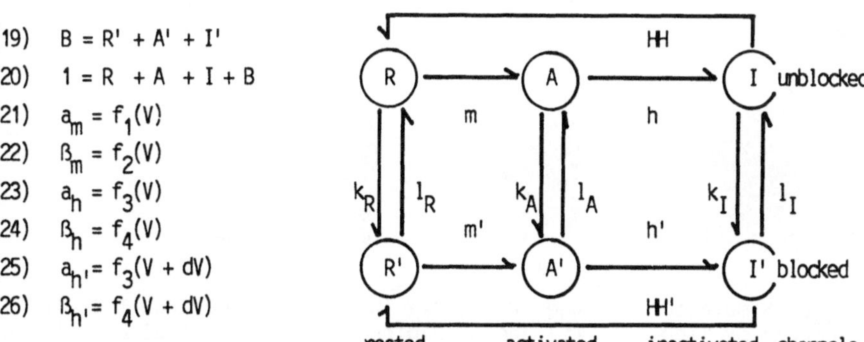

Fig.1: Scheme and model equations of the modulated receptor hypothesis for
the specific interactions of antiarrhythmic drugs with the sodium channels
in their different states of the heart cell membrane. D drug concentration;
B blocked channels; m,h,h' gating variables; a,ß rate constants; V membrane
potential; HH, HH' Hodgkin-Huxley kinetics.

Experimentally induced channel activations at different frequencies lead to
characteristic patterns of channel blockade which can be simulated using
both model structures, the BR-model and the HK-model, simultaneously. In
the following section some examples are presented:
The effects of a sodium channel blockade were simulated for the two classical antiarrhythmic drugs lidocaine and quinidine. The following figures
show the sodium currents elicited upon activation by a depolarising stimulus. It was found, that the current amplitudes are reduced with the number
of activations under the influence of the drug. This can be explained by
the fact, that during each turnover of the three channel states, a fraction
of channels will be blocked by the substance, diminishing the number of con-
ducting channels. Sodium current amplitude is continuously reduced until

157

the number of channels being blocked during one activation equals the
number of channels from which the drug dissociates during the same period.
The steady state amplitude of the sodium current therefore strongly depends
upon the activation frequency.

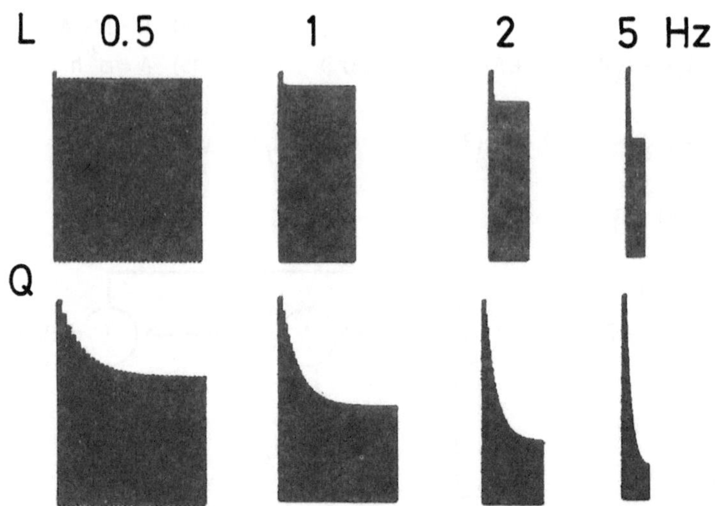

Fig.2: Simulations of the frequency- (0.5 - 5 Hz) dependent sodium channel
blockade by lidocaine (L) and quinidine (Q). The vertical bars indicate the
maximal amplitude of I_{Na} during repetitive membrane stimulations.

Fig.2 demonstrates the influence of different frequencies upon sodium channel blockade by lidocaine and quinidine. Obviously, the onset of block is
accelerated with increasing frequencies and blockade consecutively reaches
higher levels. The difference in the actions of lidocaine and quinidine is,
that for quinidine it takes a longer time to reach its steady state blockade, whereas the effects of lidocaine are quickly attained.
In fig.3 results are shown demonstrating the recovery of the sodium channel
blockade. After onset and steady state of the drug effect during repetitive
activation a variable period of rest followed by a single test activation
was applied to the preparation. During rest drug molecules may dissociate
from I'-channels and subsequent recovery via I -> R takes place without
further formation of drug-receptor-complexes. The amplitude of the sodium
current upon test activation therefore increases in a characteristic, i.e.
exponential, way. It is demonstrated that lidocaine enables quicker recovery, whereas under quinidine channel blockade recovers slowly. The recovery
kinetic of propafenone is also rather slow, but in contrast to quinidine,

158

shows only incomplete recovery. Explanations of these effects can be given
in terms of the interaction parameters found for the three drugs (tab.1).

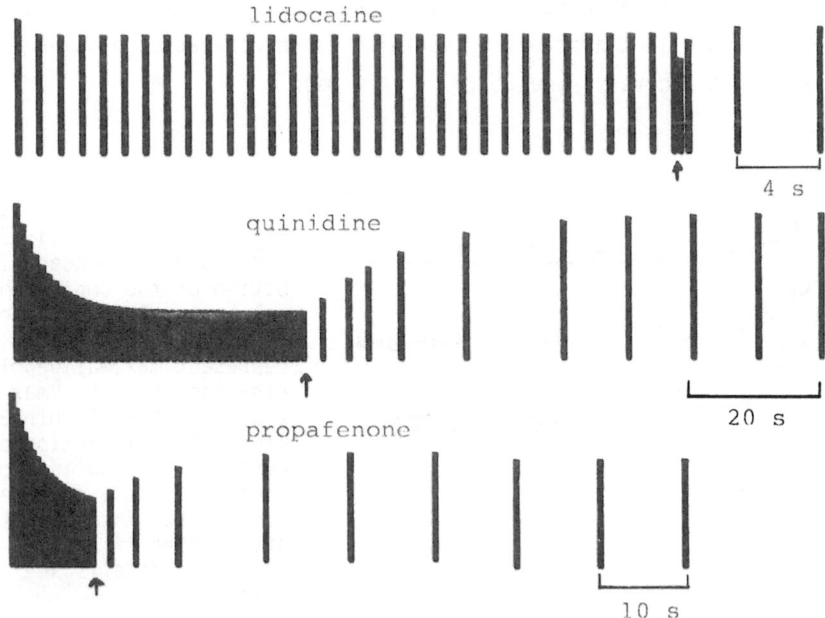

Fig.3: Simulations showing onset of sodium channel blockade and recovery
from block under the influence of lidocaine, quinidine and propafenone. Vertical bars indicated the maximal amplitude of I_{Na}. For the method see text.
Begin of recovery: ↑

These parameters have been evaluated by different authors (3,4) using
global parameter estimation procedures, simultaneously taking into account
all the experimental results available. Characteristic differences have
been found for the three substances. Quinidine possesses a moderately high
affinity only for activated channels, its time constant τ_I = 16 s for the
dissocation from inactivated I'-channels being very high. In contrast, lidocaine shows a high affinity for activated and inactivated channels, its τ_I
= 280 ms is very small. These differences explain the slow kinetics of
blockade and recovery of quinidine and the fast kinetics shown for lidocaine. Propafenone is a pure inactivation blocker (i.e. with an affinity to
inactivated channels only), which is the reason for its incomplete recovery.
Fig.4 shows our own experimental results concerning the frequency-dependent
sodium channel blockade by the new antiarrhythmic drug tocainide in guinea
pig left atria. The data obtained from experiments with stimulation frequen-

cies of 1, 2 and 3.3 Hz were used for a global parameter estimation. A reasonable fit could be achieved in spite of the fact, that there was no information about the voltage shift of h', which was set to a value of 50 mV. The parameters (tab.1) characterize tocainide as an activation blocker with a moderately fast recovery kinetic (τ_I = 1420 ms).

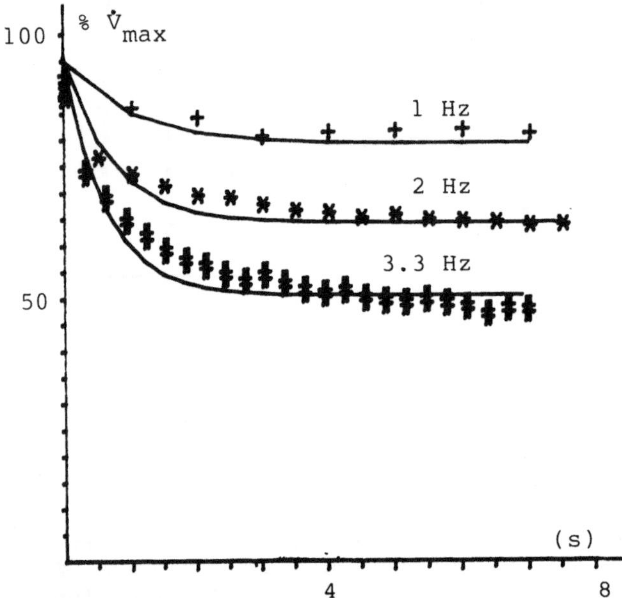

Fig.4: Frequency- (1, 2 and 3.3 Hz) dependent inhibition of the sodium channel by tocainide. The experimental values (+,*,#) represent maximal depolarisation velocity $\dot{V}max$ (as % of control) of guinea pig left atria action potentials. The solid lines correspond to the fitted model data. Abscissa: time after onset of stimulation. For details see text.

In order to apply the hypothesis of the modulated receptor to results obtained from our experiments with the calcium channel antagonists diltiazem and nifedipine the original model was formulated for the calcium channel. The experiments were performed with slow response action potentials, which are elicited under elevated extracellular potassium concentration. Under these conditions resting potential of the membrane is about V_m = -50 mV, thus inactivating the fast sodium inward current. The slow response then will be carried mainly by the calcium inward current. Its maximal depolarisation velocity $\dot{V}_{max}$ can be regarded as a measure of the calcium current. In fig.5 the onset of calcium channel block, as observed in experiments with guinea pig papillary muscles, is compared to corresponding simulation studies. Two different drug concentrations (1 and 3 µmol/l) were used. In these simulations it was not intended to find the optimal fit, but rather to reproduce the major qualities of the experimental data. The model parameters are given in tab.1. In contrast to diltiazem, which attains its steady state blockade after about 20 activations, nifedipine (fig.6) shows a different pattern of blockade: The steady state level is already reached within

the second activation. The simulation studies support the following explanation: In both cases an affinity for activated channels was assumed, leading
to the blockade of a certain number of channels per activation and a reduction of $\dot{V}_{max}$.

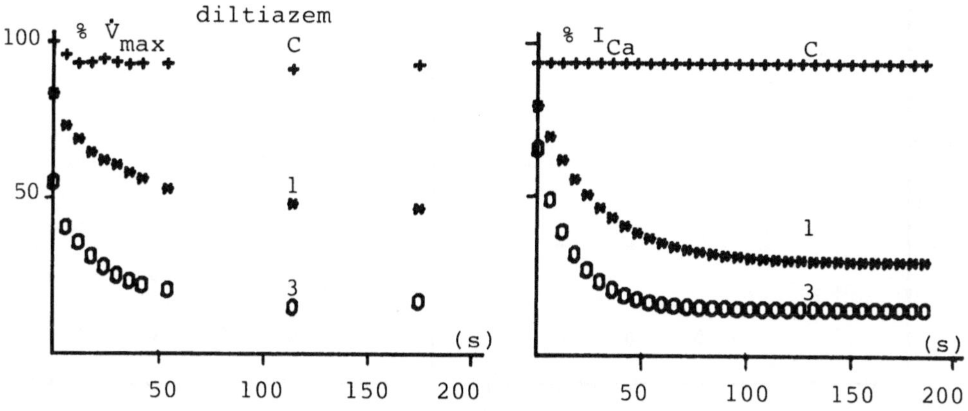

Fig.5: Concentration- (1 and 3 µmol/l) dependent inhibition of the slow
inward current I_{Ca} by the calcium antagonist diltiazem. C = control. Abscissa: Time after onset of stimulation. Left: Experimental data as maximal
depolarisation velocity Vmax (as % of control) from slow responses of
guinea pig papillary muscles. Right: Simulated model data.

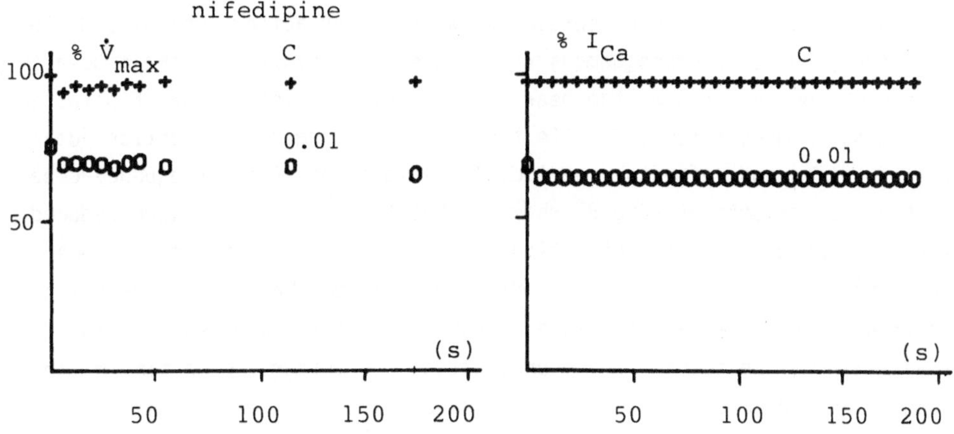

Fig.6: Inhibition of the slow inward current I_{Ca} by the calcium antagonist
nifedipine (0.01 µmol/l). Method as in fig.5.

While dissociation from inactivated channels for diltiazem takes a long
time (τ_I = 100 s), and subsequent accumulation of blocked channels may
take place, the corresponding time constant (τ_I = 1 s) for nifedipine is

small enough, such that complete recovery of blocked channels occurs between activations.
Recovery from diltiazem induced channel block is shown in fig.7. The same
model parameters (tab.1) were used as for the simulation in fig.5 resulting
in a reasonable quality of fit.

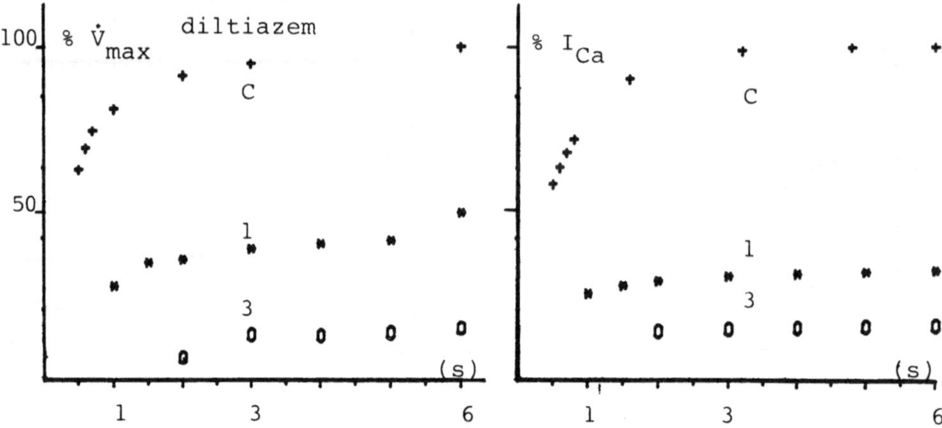

<u>Fig.7</u>: Recovery from calcium channel blockade under different concentra-
tions (1 and 3 µmol/l) of diltiazem as compared to control (C) values. Abscissa: duration of test interval. Left: Experimental data as Vmax values obtained from slow responses of the guinea pig papillary muscle. Right: Simulated model data.

The new antiarrhythmic drug sotalol shows a way of action, which is different from the other drugs demonstrated here. Sotalol reduces the repolari-
sing membrane currents of the heart muscle cells by inhibition of potassium
channels. In myocardium this effect causes a prolongation of action potential duration and refractory period. Investigations of the frequency-dependent action of sotalol (fig.8) showed, that the drug effects were reduced
with increasing frequency of activation. At the first sight, this seemed to
be opposite to the results with sodium and calcium channel antagonists as
their effects increase with higher frequencies ("use dependence"). The con-
tradiction may be solved if one understands the integral open time of the
channel as the major determinant for the use dependent drug effect. The
open time of the K-channel in turn is well correlated with action potential
duration. Action potential duration decreases with increasing stimulation
frequencies under physiological conditions, thus reducing the available interaction time for the drug with open channels. This interpretation was
tested by application of the modulated receptor hypothesis to the K-channel.
For this purpose the model equations were formulated for the K-channel and

combined with the BR-model for the simulation of action potentials. Action
potentials with durations as observed in our experiments in the frequency
range of 1/min to 120/min were simulated. In the simulation sotalol prolonged action potential duration measured at the level of 90% repolarisation
(APD_{90}). The prolonged APD was evaluated in percent of the drug-free control and plotted against the stimulation frequency.

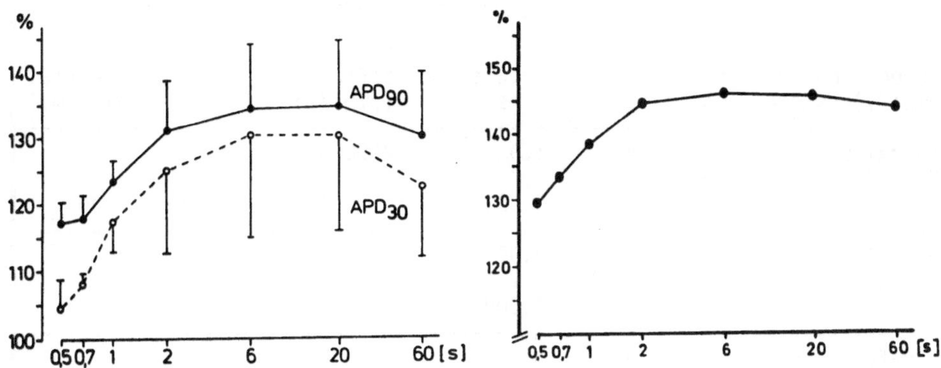

<u>Fig.8</u>: Frequency dependent broadening of action potentials from guinea pig
papillary muscle under the influence of the K-channel inhibitor sotalol.
Abscissa: interval between repetitive stimuli. Left: Experimental data showing increase in action potential duration at 30 and 90 % repolarisation
compared to control values. Right: Simulated model data for APD_{90} as described in the text.

In fig.8 the simulated results show a good fit to the experimental data,
the simulated APD values being over-estimated by an average value of 10%.
This discrepancy could certainly be reduced by further optimization of the
model parameters, but it was our aim to reproduce the experimental finding,
that sotalol shows a use dependent effect in the sense described above. The
conclusion is, that the modulated receptor hypothesis could be applied
equally well to the K-channels and to K-channel blocking drugs, respectively.
In this context, it would be of interest to investigate the dynamic effects
of the drug, as it has been shown for Na- and Ca-channel antagonists. We
had to restrict our studies to steady state effects of sotalol, because the
BR-model does not reproduce well enough the relation between stimulation
frequency and action potential duration. As the APD seems to be the major
determinant for the drug effect, a reasonable simulation of the dynamics
could not be achieved.

Recently, a new model for the reproduction of the electrical activities in
the heart cell membrane was published (5). The model contains additional
components for the description of dynamic concentration changes in the
intra- and extracellular space, concerning the most important ions as Na,
Ca, and K. Furthermore, active ion transport systems as Na-K-pump and
Na-Ca-exchange have been incorporated. It seems possible, that further studies with sotalol could be performed on the basis of this model.

Table 1.

drug (μmol/l)	channel state	association (ms mol/l)$^{-1}$	dissociation ms^{-1}	affinity (μmol/l)	τ_T ms	potential-shift(mV)
lidocaine	R	0.4	1.0	2500000.		
30 - 60	A	50000.	50.	30.		
	I	50.	0.002	40.	280	30
quinidine	R	0.	0.05			
15 - 30	A	26000.	2.7	103.		
	I	0.	0.00006		16000	40
propafe-	R	3.3	1.	300000.		
none	A	1.	1.	1000000.		
20	I	163.	0.000094	0.6	298	55
tocainide	R	0.	14.38			
200	A	58800.	2.48	421.		
	I	0.00564	0.0007	120000.	1420	50
diltiazem	R	0.	0.05			
1 - 3	A	8960.	0.00896	0.1		
	I	0.	0.00001		100000	55
nifedipine	R	0.	0.05			
0.01	A	2688170.	0.000896	0.0003		
	I	0.	0.001		1000	55
sotalol	R	0.4	1.	2500000.		
100	A	20.8	0.06	2900.		

Literature.
1. Beeler GW, Reuter H (1977) J Physiol 268:177-210
2. Hodgkin AL, Huxley AF (1952) J Physiol 117:500-544
3. Hondeghem LM, Katzung BG (1977) Biochim Biophys Acta 472:373-398
4. Kohlhardt M, Seifert C, Hondeghem LM (1983) Pflügers Arch 396:199-209
5. DiFrancesco D, Noble D (1985) Phil Trans R Soc Lond B 307:353-398

Simulation of the Human Blood Circulatory System with the help of an Uncontrolled Pulsatile Model and its Validation

Thomas Sikora, Berlin
Dietmar P.F. Möller, Mainz
Vaclav Pohl, Bremen
Ewald Hennig, Berlin

Summary. Results of the simulation of the human circulatory system by an uncontrolled pulsatile model are demonstrated. The model allows studies of the orthostatic and pathophysiological states of the circulatory system. It could be proven that the results are in qualitative good agreement with published human data. The model can be used for computer simulation of control systems of the total artificial heart and their comparative testing and their optimization.

Zusammenfassung. In der vorliegenden Arbeit werden Simulationsstudien an einem ungeregelten pulsatilen Modell des Kreislaufsystems vorgestellt. Mit Hilfe dieses Modells können sowohl orthologische als auch pathologische Kreislaufzustände nachgebildet werden. Die erzielten Simulationsergebnisse befinden in guter Übereinstimmung mit bekannten Daten aus der Literatur. Damit kann das Modell auch für Simulationsstudien zum künstlichen Herzen und hierbei für den Entwurf optimaler Regler eingesetzt werden.

1. Introduction

The computer simulation is a good tool to replace or reduce the necessity of animal experiments for the investigation of circulatory assist and replacement devices. The mathematical models of the cardiovascular system [2,3] are very costefficient tools and they allow the fast imitation of a great number of variations even in extreme pathophysiological cases. On the basis of the model described in [1] a new model has been developed that takes the pulsatile behaviour of the circulation into consideration.

The contents of information of this model is much higher because the pulsatile pressure changes and the resulting changes in blood volume and flow in the ventricles are much more realistically.

With the help of these models control hypotheses for the control system, necessary for artificial heart assist and replacement devices, can be studied in vitro. Existing control concepts can be optimized with the very close simulation of the complexity of the cardiovascular system. Extreme pathophysiological states e.g. heart valve stenosis and heart valve insufficiencies can be described realistically with the pulsatile model and taken into consideration of the control mode of the artificial heart.

2. Description of the mathematical simulation model

The basic cardiovascular model described in [1] has been extended with two cardial components for the left and right ventricle including the 4 heart valves.
Figure 1 shows the structure of the uncontrolled physiological closed pulsatile model
of the 6th order. It consists out of the
following main components derived from hydromechanical modelling.

- 2 cardial, capacitive compartments KKL(t)
  KKR(t) for the left and right ventricle
  where the elasticity of the ventricles was
  assumed to be timevariant and to be the
  driving forces for the pressure changes
  during the systoly.
- 4 resistive valve compartments KAK, KMK,
  KTK, KPK for the simulation of heart valve
  functions. The flow resistance has been
  described by the pressure difference on
  both sides of the respective valve.
- 2 arterial, elastic compartments CAS and CAP describing the
  systemic and pulmonary arterial circulation.
- 2 venous, capacitive compartments CVS and CVP.

The model is described by a set of non-linear timevariant differential equations of first order which can be solved by computer simulation. The normal pumping function of the natural ventricles with the opening and closing of the inlet or outlet valves according to the momentarily acting pressure

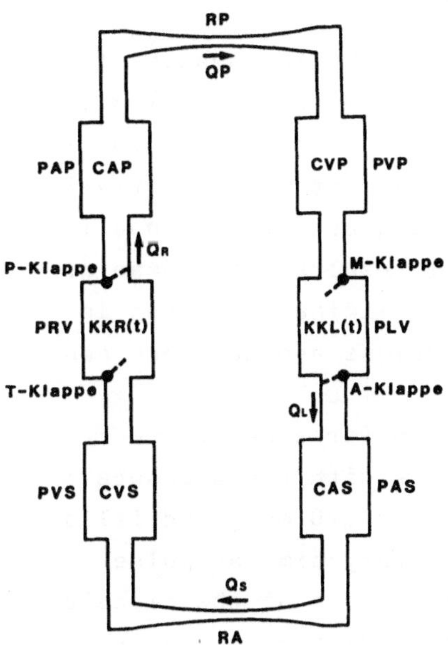

Fig. 1 Hydromechanical structure of the circulation model

differences is very well simulated by the model. The pressure
changes in the ventricles with closed valves during the contraction phase followed by an opening of the aortic and pulmonary artery valves if the necessary pressure difference has
been reached leads to a resulting increase in pressure in the
aorta and the pulmonary arteries with a curvature very close
to the natural pulse contour. In the case of decreasing pressure in the ventricles during diastoly the aortic pressure
decreases simultaneously because of the volume compensation
between the arterial and the venous compartment. If the ventricular pressures are decreasing below the pressures in the
venous parts, the atrial ventricular valves are now opening
and the ventricles will be filled again with blood and new
systoly is iniciated.

The amount of pressure PLV, PAS, PVS, PRV, PAP and PVP in
the different compartments depends on the preset time-constants of the system. The developed non-linear mathematical
model is based on a structure of the real biological system
with its functional and morphological attributes. The model
parameters are therefore biologically interpreted.

3. Results

The demands for a circulatory model allowing the design of
optimal controllers are very complex. The model must be able
to describe pulse contours and changes in blood volume in
different parts of the human circulatory system qualitatively
as well as quantitatively with satisfying accuracy. A qualitative comparison of the complex model results with human or
animal experimental datas is often a preliminary process in
the modelling of physiological interactions and has been done
in [4].

Figure 2 shows changes in blood pressure for a healthy person. The aortic pressure (PAS) increases with the pressure in
the left ventricle (PLV) to approximately 120 mmHg and falls
during the diastoly down to 80 mmHg. A very similar pulse
contour has been found for the pulmonary part of the circulatory system. The ventricular pressure (PRV) and pulmonary
artery pressure (PAP) increases to approximately 40 mmHg and
PAP decreases to an end-systolic (dia-
stolic) value of nearly 20 (10) mmHg.
A comparison with simulation results
of the pulsatile model as shown in Fig.3
shows a qualitatively good agreement.

Under pathophysiological conditions the
blood pressure changes are abnormal and
a mathematical model must be able to
simulate even such cases. Figure 4 shows
a physiological example for the changes
in pulse curvature due to aortic and
pulmonary valve stenosis. Stenotic outflow valves are forcing the ventricle
to build up a much higher end-systolic
pressure.

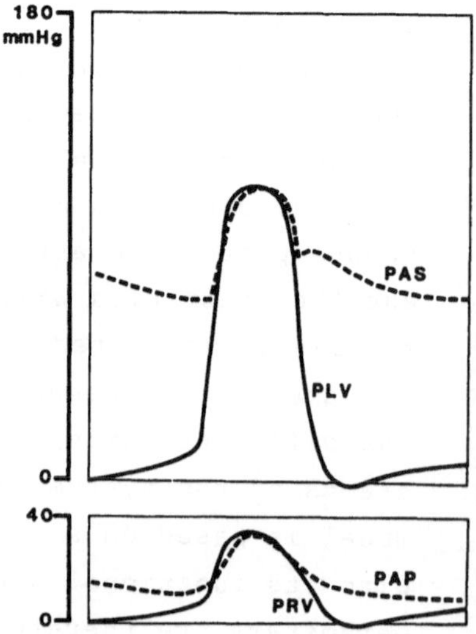

Fig. 2 Blood pressure
transients for
the normal case

The pressure losses depending on the degree of the valve stenosis are much higher, such resulting in a respective lower systolic aortic pressure with a small time delay. The mathematical model is able to reproduce this extreme pathophysiological state with very high accuracy (Figure 5).
To simulate this pathophysiological state the resistance of the valve in the model has been increased. The changes of the

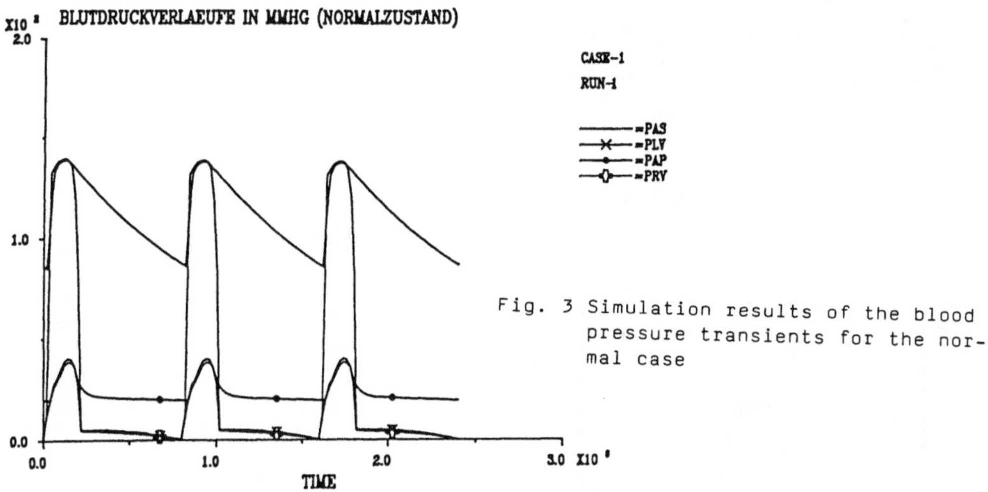

Fig. 3 Simulation results of the blood pressure transients for the normal case

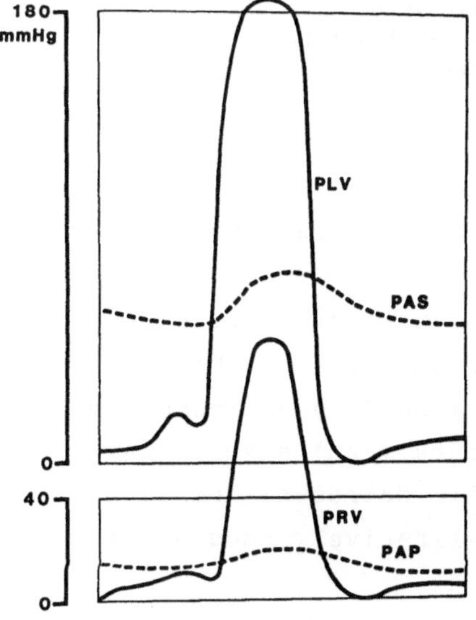

Fig. 4 Blood pressure transients under aortic and pulmonary valve stenosis

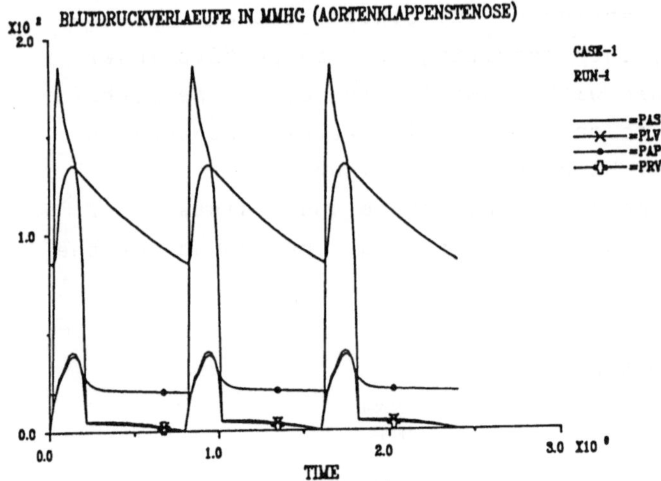

Fig. 5 Simulation results for the case
        study of the aortic valve stenosis

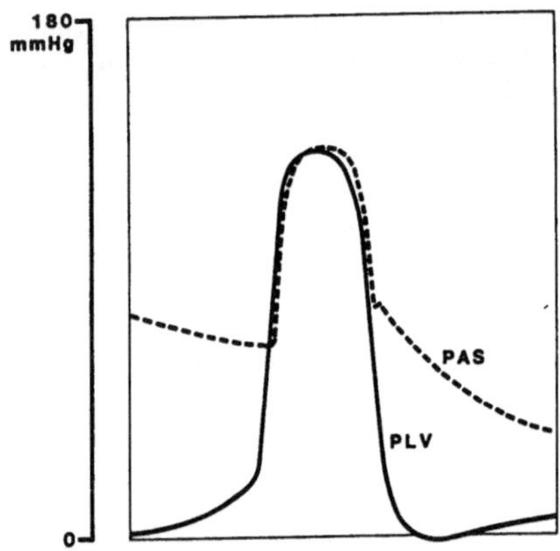

Fig. 6 Blood pressure
        transients un-
        der aortic
        valve insuffi-
        ciency

in- and outflow pressure differences are the respective simulation results.

Figure 6 demonstrates typical physiological pressure curves
in the case of aortic valve insufficiency. Due to the arterial backflow the left ventricle is exposed to an increased
volume load. The arterial blood pressure (PAS) shows high
systolic pressure with a deep diastolic decrease. The simulated results (Figure 7) are in a qualitatively good agreement with the physiological example.

170

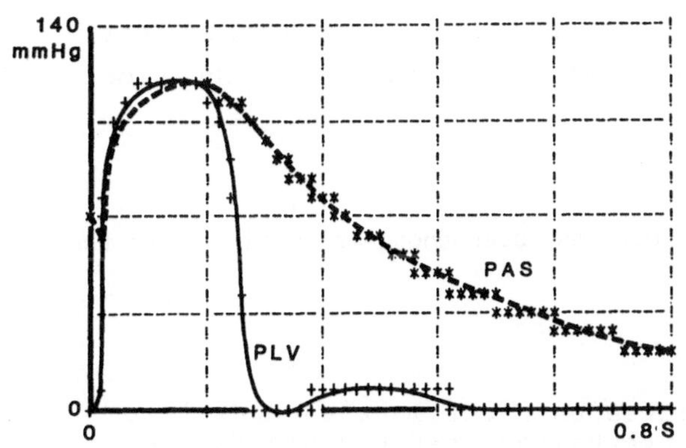

Fig. 7 Simulation re-
sults for the
case study
aortic valve
insufficiency

This relatively close coincidence of the physiological and
simulated results has proven that the model parameters and
the model structure was selected in an appropriate way. This
allows   statements about other physiological significant
parameters or curves which are not easily, or sometimes even
not at all measureable invasively in the human body. As an
example the blood flow through the aortic valve (QL) is shown
in Figure 8 as a simulated result in case of aortic valve insufficiency. Clearly demonstrated is the calculated backflow
leading to an additional volume load for the left ventricle.

Therefore the left ventricle has to eject an additional amount
of blood volume during each systoly to keep the balance with
the flow through the right
ventricle and such the flow
equilibrium through the
systemic and pulmonary circulation.

Subject of further work
will be the quantitative
comparison with human and
animal experimental datas.
Because of the fast time
variant changes in the parameters of the circulatory system a direct compari-

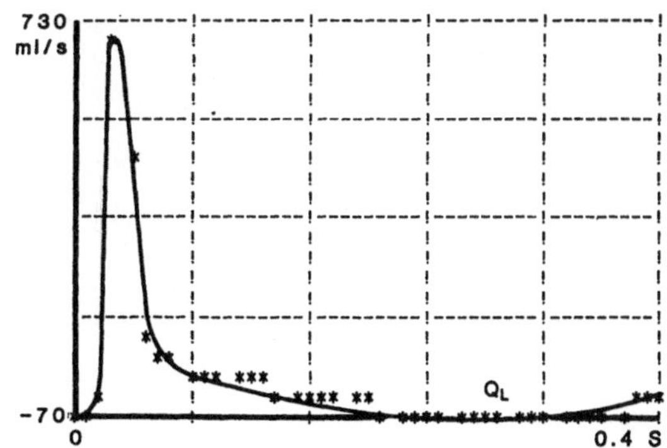

Fig. 8

Simulation results for the case study
blood flow through the insufficient
aortic valve

son is only possible with the application of parameter estimation and parameter identification procedures on the simulation model.

This work was supported by the Deutsche Forschungsgemeinschaft under the
grant He 1265/2-1.

Literatur:

/1/ Möller, D.: Ein geschlossenes nichtlineares Modell zur
    Simulation des Kurzzeitverhaltens des Kreislaufsystems
    und seine Anwendung zur Identifikation
    Reihe: Medizinische Informatik und Statistik, Bd. 30
    Springer-Verlag, Berlin-Heidelberg-New York, 1981
/2/ D. Möller, V. Pohl, T. Sikora, E. Hennig: Simulation eines
    ungeregelten pulsatilen Modells des Herz-Kreislauf-Systems
    In: Informatik Fachberichte, Bd. 109, Simulationstechnik,
    S. 346-349, Hrsg.: D. P. F. Möller
    Springer-Verlag, Berlin-Heidelberg-New York, 1985
/3/ Möller, D.: Computersimulation der renalen Hämodynamik
    In: Informatik Fachberichte, Bd. 109, S. 366-370,
    Hrsg.: D. P. F. Möller
    Springer-Verlag, Berlin-Heidelberg-New York-Tokyo, 1985
/4/ Möller, D.: Ein mathematisches Simulationsmodell zur Unter-
    suchung der Kreislaufregulation
    In: Durchblutungsregulation und Organstoffwechsel,S.190-196
    Hrsg.: J. Grote, E. Witzleb
    Akademie der Wissenschaften und Literatur, Mainz, 1982

Model-Reduction for the Parameter Identification of an Uncontrolled Pulsatile Model of the Cardiovascular System

Vaclav Pohl, Bremen
Dietmar P.F. Möller, Mainz
Thomas Sikora, Berlin
Ewald Hennig, Berlin

Summary. The paper outlines an identification scheme for the parameter identification of an uncontrolled pulsatile model of the cardiovascular system. Model-reduction is introduced by seperation of the opening and closing interval of the valves. These simplified models, depending to different phases of the heart action (systolic and diastolic), allow a faster parameter identification.

Zusammenfassung. In der vorliegenden Arbeit wird ein Identifikationsverfahren zur Schätzung der Parameter eines ungeregelten pulsatilen Kreislaufmodelles vorgestellt. Dabei werden diejenigen Öffnungs- und Schließzeitpunkte der Herzklappen ausgewählt, mit denen eine Modellreduktion möglich ist. Mit Hilfe derart vereinfachter Modelle für die einzelnen Phasen einer Herzaktion (Systole und Diastole) können die, für die Regelung relevanten Parameter schneller identifiziert werden.

1. Introduction

The dynamic behaviour of the cardiovascular system was described with sufficient accuracy by a mathematical model. With respect to the case study approach, modeling and identification depends from the following facts:

- Ascertainment of in vivo non-measurable or difficulty-measurable physiological variables,
- Ascertainment of parameters, which are proper useful for the control of the cardiovascular system.

Pulsatility was applied as an system excitation for the parameter identification method, used. The parameters, identified in this way, are estimated fast. But, this needs fast and exact find outs of the relevant physiological quantities.

2. Model-Simplification

The uncontrolled pulsatile model of the cardiovascular system [1] is a non-linear, time-varying system, of 8th order. The non-linearity is based on the heart valves, whereby the time-

varying features are derived from the valves as well as the
compliances of the left and the right ventricles.
Assuming opening and closing the valves are done within negli-
gible time, the heart-action can be divided into a systolic-
and diastolic- interval as shown in Fig. 1, with the possibility of separation into subsidary models, as shown in Fig. 2.
In comparison with a closed model the separation of the cardiovascular system yields into an easy done and fast parameter identification by means of the simplified model-structure.

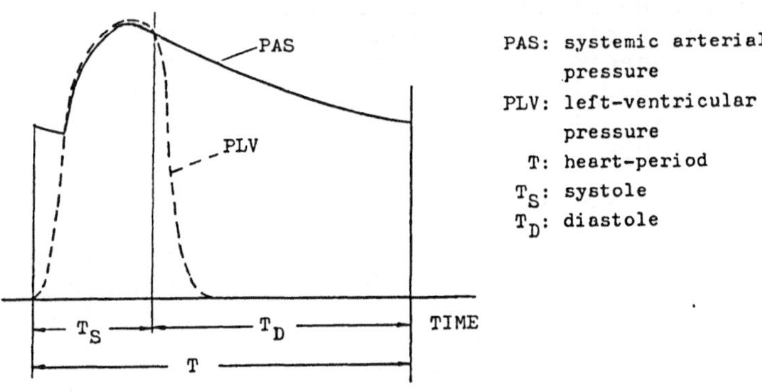

PAS: systemic arterial
     pressure
PLV: left-ventricular
     pressure
  T: heart-period
T_S: systole
T_D: diastole

Fig. 1: Schematic diagram of the blood-pressure dependencies
        under one heart-beat

Model-reduction was performed with the SLCS-4 simulation-
package. Simulation runs shows, that the results, obtained
from the simplified subsystems are identical with that, obtained from the closed model within comparable time-intervals.

3. Parameter-Identification

Identification is used to represent the measurable in-output-
behaviour of a cardiovascular system by means of a reference-
model, adjusting the parameters as precise as possible [2].
The identification of the parameters e.g. CAS, RA etc. are not
of primary interest. Because the dynamic behaviour is determined by the time-constants e.g. T_1 = CAS·(RA + 1/KVS) etc.
Therefore it is only necessary to identify the time-constants
for each sub-system.
The procedure for identification the time-constants should be
demonstrated, using as an example the diastolic-model, as
shown in Fig. [3].

174

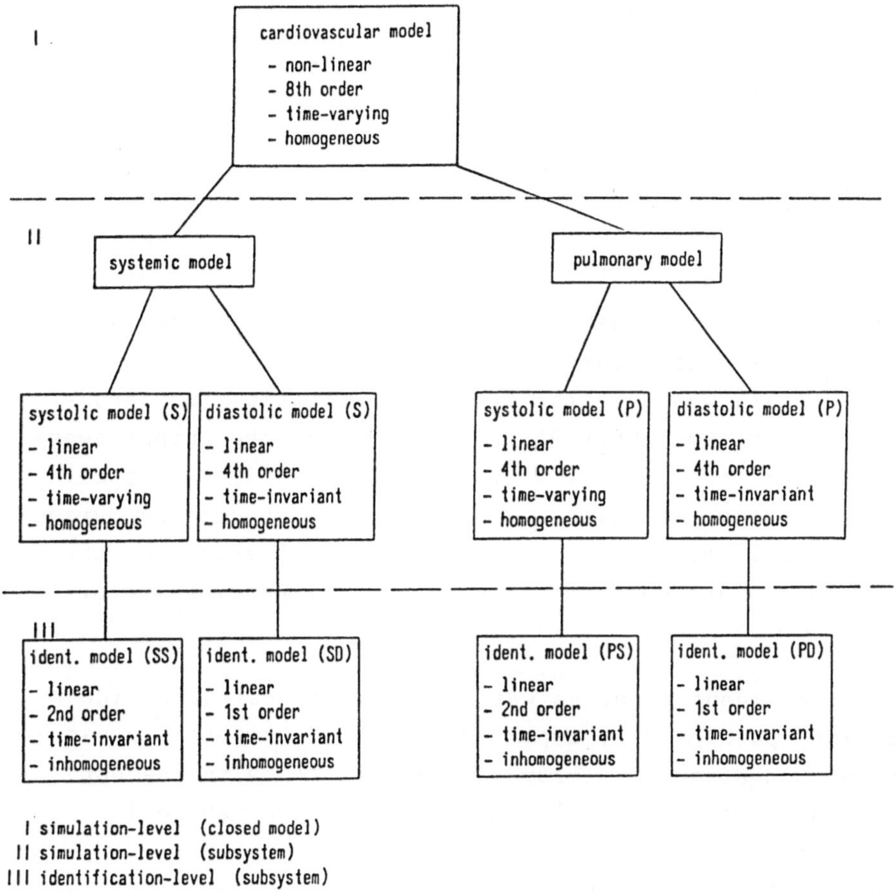

I
cardiovascular model
- non-linear
- 8th order
- time-varying
- homogeneous
II
systemic model
pulmonary model
systolic model (S)
- linear
- 4th order
- time-varying
- homogeneous
diastolic model (S)
- linear
- 4th order
- time-invariant
- homogeneous
systolic model (P)
- linear
- 4th order
- time-varying
- homogeneous
diastolic model (P)
- linear
- 4th order
- time-invariant
- homogeneous
III
ident. model (SS)
- linear
- 2nd order
- time-invariant
- inhomogeneous
ident. model (SD)
- linear
- 1st order
- time-invariant
- inhomogeneous
ident. model (PS)
- linear
- 2nd order
- time-invariant
- inhomogeneous
ident. model (PD)
- linear
- 1st order
- time-invariant
- inhomogeneous
I simulation-level (closed model)
II simulation-level (subsystem)
III identification-level (subsystem)

Fig. 2: Scheme of the model-reduction used for the application of parameter-identification

PAS
PVS
PRV
CAS
CVS
1/KKR
AK
TK
PK
RA+1/KVS
1/KRV
AK: aortic-valve
TK: tricuspidalic-valve
PK: pulmonary-valve

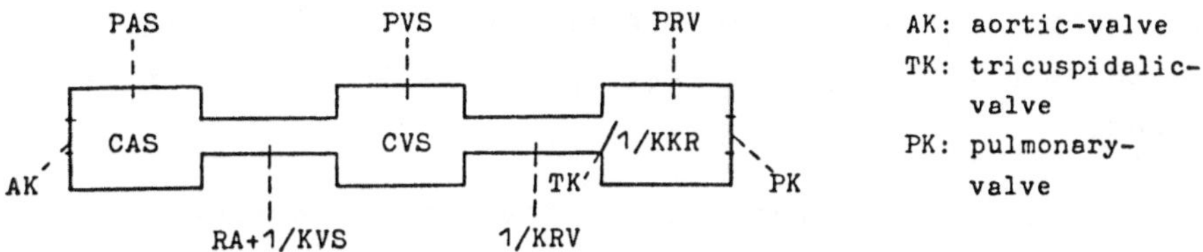

Fig. 3: Compartment-representation of the systemic subsystem during the diastolic interval, applied for the identification of the time-constants

During the diastolic interval, the aortic-valve and the pulmonary-valve are closed, while the tricuspidalic-valve is opened. The subsystem can be described by the following equations:

$$PAS = -1/T_1 \ (PAS - PVS) \qquad\qquad (1)$$

$$PVS = 1/T_2 \ (PAS - PVS) - 1/T_3 \ (PVS - PRV) \qquad (2)$$

$$PRV = 1/T_{KR}(PVS - PRV), \qquad\qquad (3)$$

with

$$T_1 = CAS \ (RA + 1/KVS) \qquad\qquad (4)$$

$$T_2 = CVS \ (RA + 1/KVS) \qquad\qquad (5)$$

$$T_3 = CVS/KRV \qquad\qquad (6)$$

$$T_{KR} = 1/(KRV \cdot KKR(t)), \qquad\qquad (7)$$

with the assumption, that KKR(t) is approximately constant during the systolic interval.

The pulsatility was used for system's excitation for parameter identification, where PAS is regarded as the output quantity. However, the initial states of PASO, PVSO, PRVO shall be known, that means certain values must be measurable during the relevant interval. The data which are necessary for the identification are obtained from a physical-cardiovascular model.

The off-line-identification was carried out, using the identification- package NLP, implemented on PCS-computer, based on the method of output-error in combination with the optimization-strategy of Powell and Rosenbrock, and shown in Fig. 4 .

The reference-model was described by equations (1-3). At first the identification was performed to identify T_1 which primary determines the dynamic behaviour of PAS during the diastolic interval. This parameter has been identified under various initial- conditions in comparatively short time ($\sim$ 15 sec). The remaining parameters (T_2, T_3, T_{KR}) were only roughly approximated.

Furthermore it could be recognized, that the parameter T_1 might be instantly accessible by the following analytical model-simplification:

$$T_1 \approx \frac{t_D}{-\ln \dfrac{PAS(t_D) - PVSO}{PASO - PVSO}} \qquad\qquad (8)$$

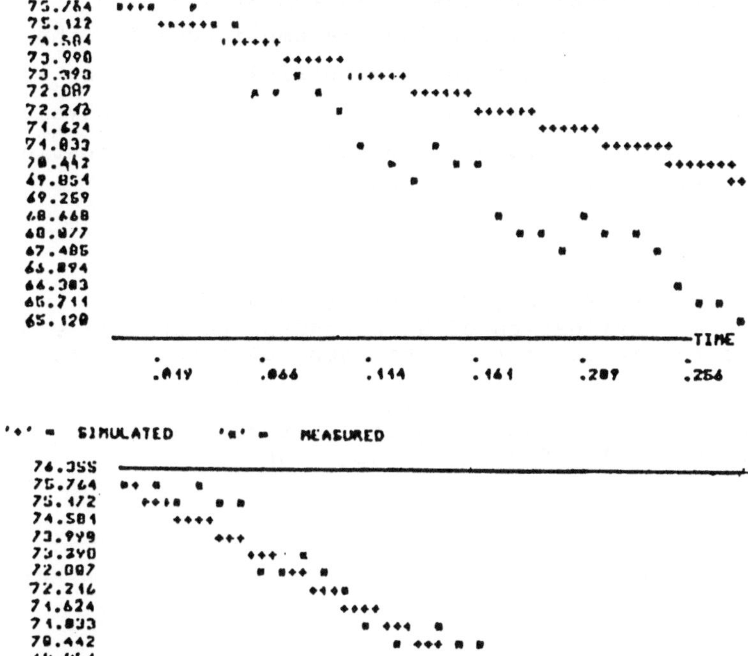

Fig. 4: Comparison of the output-behaviours before and after parameter optimization

4. <u>Conclusion</u>

It was shown, that the valve-function of the heart allows the simplification of a cardiovascular system model and speed up the identification time. The number of parameters to be identified corresponds to the time-constants of the pulsatile model, used. The inquiry of important parameters were performed by means of the parameter identification during the diastolic-period. Further investigations with this reduced order model for the identification during the systolic-period, still going on in work.

Some difficulties arise, when investigate the venous-part of
the cardiovascular system, because this requires a high accuracy in measurement due to low pulsatile amplitudes for the
identification. However, it should be possible to find an appropriate sample-schedule by means of a sensitivity-analysis
leading to improved and more precise parameter-identification.
This item is a field of further research work.

5. References

[1] T. Sikora, D.P.F. Möller, V. Pohl, E. Hennig:
    Simulation and validation of an uncontrolled pulsatile mo-
    del of the cardiovascular system (published in this edi-
    tion)
[2] D.P.F. Möller, D. Popovič, G. Thiele:
    Modelling, Simulation and Parameter-Estimation of the
    Human Cardiovascular System, Vieweg, 1983.

This work was supported by the Deutsche Forschungsgemeinschaft
under the grant He 1265/2-1.

On the improved Estimation of the Compliance-Parameters of the Physiologically Closed Cardiovascular System

A. Tanha*), H. Maftoon*), G. Thiele*), D. Möller**), D. Popović*)

*) University of Bremen, **) University of Mainz

Summary. The influence of the type of the sampling-schedule and of the system-excitation with respect to identifiability and estimation-accuracy of the compliance-parameters of a nonlinear model of the physiologically closed cardiovascular system will be investigated. Using the sensitivity-functions of the arterial systemic pressure with respect to the compliance-parameters it will be shown how to select the sample-schedule and system-excitation in order to get optimal estimation-accuracy.

Zusammenfassung. Der Einfluß der Art des Abtastschemas und der Systemerregung auf die Identifizierbarkeit und die Schätzgenauigkeit der Komplianz-Parameter eines nichtlinearen Modells des geregelten physiologisch geschlossenen kardiovaskulären Systems wird untersucht. An Hand der Empfindlichkeitsfunktionen des arteriell systemischen Blutdrucks gegenüber den Komplianz-Parametern wird gezeigt, wie das Abtastschema und die Systemerregung gewählt werden können, um eine möglichst große Schätzgenauigkeit zu erreichen.

1. Introduction

A significant problem in the process of modelbuilding of dynamical systems is the determination of unknown and not directly measurable system-parameters. This is especially true for biomedical systems, when in vivo-measurements of system-parameters are not possible or not tolerable, for instance in the case of the compliances of the vessels of the cardiovascular system to be investigated in this paper for diagnostic purposes with respect to, e.g. arterio-scleriosis and high blood-pressure. Therefore, well-known methods will be used here to determine indirectly the compliance-parameters by parameter-estimation on the basis of measured system-inputs and system-outputs.

The relevance of the parameter-estimates depends, of course, on the achievable estimation-accuracy /THIE84, TANH85/, e.g. significant variances of the estimates can result in totally wrong diagnostic results. However, if this is the case, the estimation accuracy can be improved by a suitable choice of the experiment. It will be shown that an optimization of the estimation-errors is possible with respect to the sampling-schedule using the sensitivity-functions of the arterial pressure relative to the

parameters to be estimated, so that further improvements are only possible
by reducing the errors of the blood-pressure measurements.

2. Mathematical model of the cardiovascular system

The regulated cardiovascular system will mathematically be described by a
model comprising the systemic and the pulmonary circulation by an arterial
pure elastic or capacitive, respectively, a pure resistive and a venous pure
capacitive compartment /MöLL81/. This results in a mathematical model consisting of a set of ordinary nonlinear state differential equations of
n-th order

$$\dot{\underline{x}} = \underline{f}(\underline{x},\underline{u},z,\underline{\theta}_s), \tag{1a}$$

where the first four components of $\underline{x}$ are the mean arterial and venous
pressures. The regulation of the arterial systemic pressure PAS($=x_1$) in the
baroreceptor   feedback loop will be mathematically described by

$$\underline{u} = \underline{g}(x_1) , \tag{1b}$$

with heart-frequency HF and peripheral resistance RA as components of $\underline{u}$.
Furthermore, in (1a), z specifies the system-exatation by an ergometric
workload and $\underline{\theta}_s$ is the parameter-vector to be identified, comprising the
components CASN, KCVS,KCAP, KCVP /MöLL82/.

3.   Parameter estimation

3.1 Parameter estimation method

In the method of output-error least-squares, used here, the vector of
parameter estimates $\hat{\underline{\theta}}$ is determined as the minimizing argument of the
error-functional

$$J(\hat{\underline{\theta}}) = \sum_{K=1}^{N} (y_{Meß,k} - \hat{y}_k(\hat{\underline{\theta}}))^2. \tag{2}$$

$y_{Meß,k}$ is the measured system-output and y_k the output of the identifi-
cation-model /MöLL83 , THIE84/.

3.2 Identifiability

A necessary condition for consistent parameter-estimation from noise-
corrupted data is the possibility to uniquely identify the parameters from
noise-free true-model simulation-data /THIE84/. If the latter is true,the
system will be called identifiable .Algebraic methods for testing identifiability are not applicable with·respect to the complexity of the model
to be investigated here. Therefore, it will be tested, whether the iden-
tification-result for a large number of starting-values $\hat{\underline{\theta}}_0$, taken from a
neighbourhood of $\underline{\theta}_s$, will be an estimate $\hat{\underline{\theta}}_{min}$, whereby

$$\hat{\underline{\theta}}_{min} = \underline{\theta}_s \qquad\qquad (3)$$

with a "sufficient accuracy". This philosophy is illustrated in Fig. 1
/TANH85/.

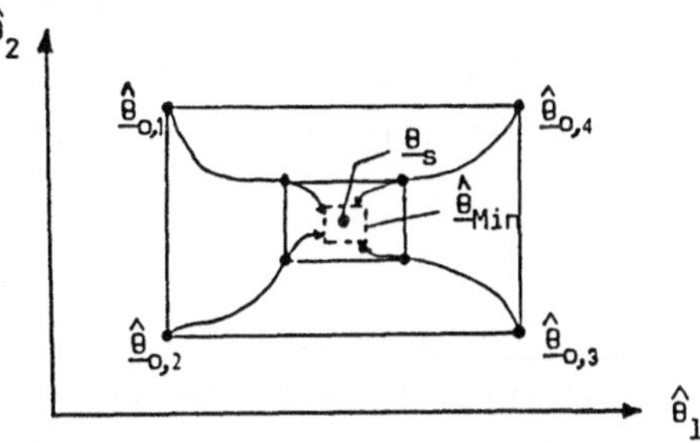

Fig. 1 Graphical representation of "testing identifiability by
identification"

3.3 Variances of estimates and sensitivity functions

As a measure of the parameter-estimation error the deviation $\Delta\hat{\underline{\theta}}_{min}$ of the
minimum of the error-functional (2) with respect to noise-corrupted
measurements $y_{Meß,k}$ in comparison to noise-free data $y_{Meß,k}=y_k$ will be
chosen. For small noise amplitudes it can be shown that /THIE84,TANH85/

$$cov\{\Delta\hat{\underline{\theta}}_{Min}\} \approx \left[(S_{\hat{\underline{\theta}}}^{\hat{y}}(\underline{\theta}_s))^T \; S_{\hat{\underline{\theta}}}^{\hat{y}}(\underline{\theta}_s) \right]^{-1} \sigma_v^2 \qquad\qquad (4)$$

where
$$S_{\hat{\underline{\theta}}}^{\hat{y}}(\underline{\theta}_s) := \left[\frac{\partial \hat{\underline{y}}^T(\underline{\theta}_s)}{\partial \hat{\underline{\theta}}} \right]^T \qquad\qquad (5)$$

is defined as the sensitivity-matrix of the vector of the model-output
samples $\hat{\underline{y}}(\hat{\underline{\theta}})$ with respect to $\hat{\underline{\theta}}$ and where σ_v^2 is the variance of the
measurement-noise.

From (4) it can be seen that the covariance-matrix of the estimation-
errors of the parameters can become significant in the case of almost
linear dependences of the columns of the sensitivity matrix. On the other
hand linear independence of the sensitivity-functions

$$s_{\hat{\theta}_i}^{\hat{y}(t)} := \frac{\partial \hat{y}(t)}{\partial \hat{\theta}_i} \qquad\qquad (6)$$

is only a necessary condition for these columns to be linear independent.
Because the i-th column comprises only the sample values $\partial\hat{y}(t_i)/\partial\hat{\theta}_i$ of the
corresponding sensitivity function, sufficiency of linear independence
has to be guaranteed by suitably choosing the sampling-schedule. This can
be done, especially, in the sense of minimization of the variances of the

estimation-errors.

If the sensitivity-functions tend to stationary values for a given system-excitation the sensitivity-matrix can become ill-conditioned too. Therefore, a suitable modification of the system-excitation, e.g. by application of an alternating input instead of step-input, there is further possibility to improve the condition of the sensitivity-matrix and, consequently, to decrease the elements of (5).

4.   Results

4.1   Identifiability of compliance parameters

Using the sample-schedule AS-I (Fig. 2) it can be shown that a constant system excitation of 100 W applied during the inverval [0s-100s] is sufficient for simultaneous identifiability of all subsets of 3 components of the parameter-vector

$$\underline{\theta} = [CASN', KCVS, KCAP, KCVP]^T$$

even in the case of single-precision arithmetic.

Simultaneous identifiability of all 4 components of $\underline{\theta}$ can only be shown, in general, by using double-precision arithmetic /TANH85/.

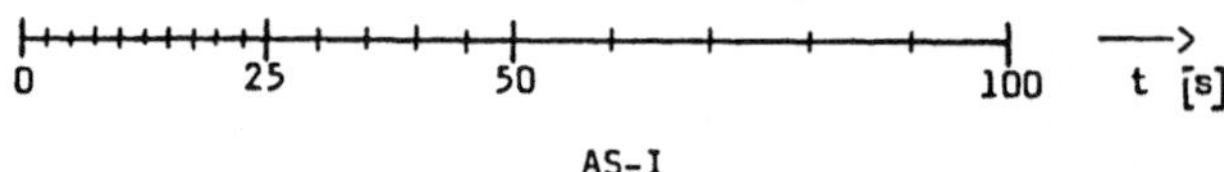

<u>Fig. 2</u> Distribution of sampling points in sampling-schedule AS-I.

Taken into account experiences with a model of the nonregulated cardio-vascular system /MÖLL85/ an improvement of the identifiability can be conjectured when using additional sample values from the subintervals where the sensitivity-functions are essentially different from zero. By using 25 additional sampling-points in the interval [0 s-2.5s] with respect to samp-ling-schedule AS-I (i.e. using the modified sampling-schedule AS-II) simultaneous identifiability of all 4 compliance-parameters can be shown even with single-precision arithmetic (Table 1). Therefore, it will be appropriate using sample-schedule AS-II in all further investigations in this paper.

4.2 Accuracy of parameter estimates

A. Influence of the type of sampling schedule

Table 2 shows approximate standard-deviations, calculated with respect to (4), for different combinations of parameters and for sampling-schedule AS-II. The approximations agree with corresponding identification results

$\underline{\theta}$	$\hat{\underline{\theta}}_o$	$\hat{\underline{\theta}}_{Min}$	$\underline{\theta}_s$	$J(\hat{\underline{\theta}}_{Min})$	Sim.
CASN	2,0	1,5	1,5		
KCVS	340,0	371,26	371,25		
KCAP	38,0	34,362	34,361	0,6647E-7	513
KCVP	47,0	43,489	43,486		

Table 1   Simultaneous identification of the four compliance-parameters with respect to sampling schedule AS-II using single-precision arithmetic.

$\mathcal{G}_{CASN}$	$\mathcal{G}_{KCVS}$	$\mathcal{G}_{KCAP}$	$\mathcal{G}_{KCVP}$
0,049 (3,2 %)	18,7 (5,0 %)	3,50 (10,0%)	3,10 (7,2 %)
0,023 (1,5 %)	9,10 (2,5 %)	1,70 (5,1 %)	-
0,023 (1,5 %)	7,40 (2,0 %)	-	-
0,021 (1,4 %)	-	-	-

Table 2   Approximated standard-deviations of the compliance parameter estimates with respect to sampling-schedule AS-II, measurement errors equally distributed over the inverval $\pm$o.o5mmHg ($\sigma_V^2 = 10^{-3}$), and ergometric workload EW=100 W.

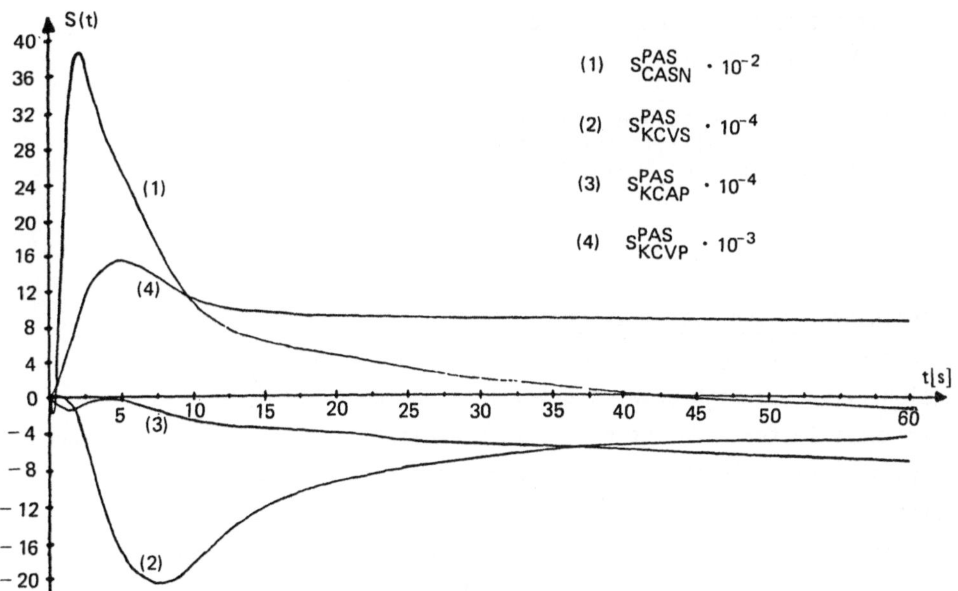

Fig. 3   Sensitivity functions of PAS with respect to the four compliance parameters and an ergometric workload EW=100 W.

corresponding identification results in the order of magnitude, but they
are, clearly, still too big to be used for physiological inferences about
the system itself. In order to decrease the variances by modifications of
the sampling-schedule, the sensitivity functions have to be computed
(Fig. 3).

It becomes evident from Fig. 3 that only a few sampling points are placed
in the subinterval where s_{CASN}^{PAS} is essentially different form zero.

Conjecturing from Fig. 3 that equidistant sample intervals of 0.1s length
will cover the interval [0s-20s] of largest values of the sensitivity functions sufficiently dense to avoid linear dependences, the variances are
computed for this case (AS-III). We find that the standard deviations
can be reduced in this way up to 50% with respect to sampling-
schedule AS-II (Table 3).

G_{CASN}	G_{KCVS}	G_{KCAP}	G_{KCVP}
0,042 (2,8 %)	9,00 (2,4 %)	1,68 (4,9 %)	1,75 (4,0 %)
0,018 (1,2 %)	3,42 (0,9 %)	1,41 (4,1 %)	-
0,017 (1,17%)	2,21 (0,6 %)	-	-
0,012 (0,8 %)	-	-	-

Table 3 Approximated standard-deviations with respect to sampling
schedule AS-III,measurement-errors equally distributed over
$\pm$ 0.05 mmHg and EW=100 W.

B. Influence of the system excitation

By inspection of Fig. 3 we find that the sensitivity-functions tend to
stationary values with growing time. This can be avoided, e.g., by using a
square-wave instead of a step as system-excitation (Fig. 4). The variance-
analysis confirms in this case a further reduction of the standard-
deviations up to 50%. Increasing the ergometric workload to 200 W in an
otherwise unchanged experiment we find a further reduction of standard-
deviations, ranging now between 1.0% and 2.4%.

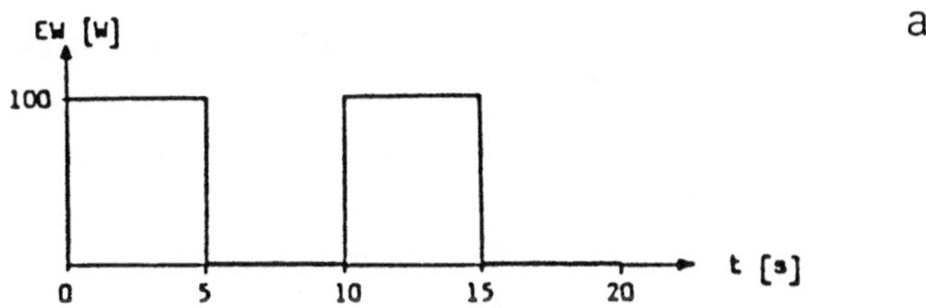

b

G_{CASN}	G_{KCVS}	G_{KCAP}	G_{KCVS}
0,024 (1,6 %)	7,20 (1,9 %)	1,60 (4,7 %)	1,00 (2,3 %)
0,009 (0,6 %)	2,14 (0,6 %)	1,50 (4,4 %)	-
0,008 (0,5 %)	2,12 (0,57%)	-	-
0,008 (0,5 %)	-	-	-

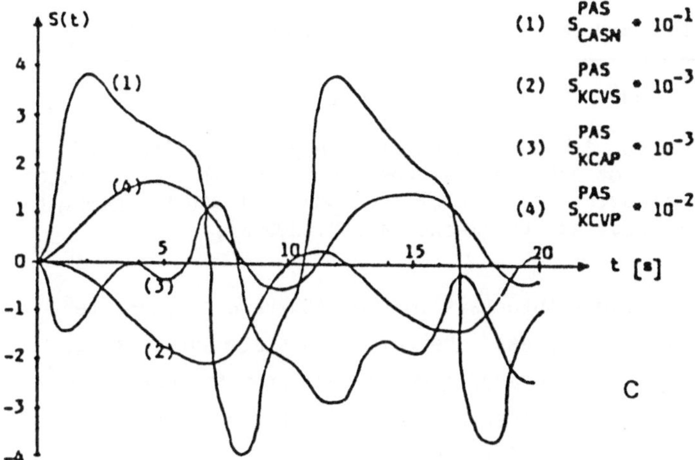

<u>Fig. 4</u> Sensitivity-functions (c) and standard-deviations (b) with respect to a square-wave as system-excitation(100 W) (a) and equally distributed measurement-erros ($\pm$ 0.05mmHg).

4.3 Discussion

With regard to application of compliance-estimates for detection of a possible arterio-sclerosis of a patient it is necessary to guarantee iden-tifiability and sufficient accuracy of the parameter-estimates with respect to the complete range of compliance-values characterizing the range of possible degrees of the illness. With reference to /MÖLL82/ we will assume that the CASN-values 1.5, 0.96 and 0.62 correspond to illness-de-

grees I, II and III of arterio-sclerosis. To be able to discriminate these
cases sufficienty well the necessary estimation-accuracy can be characterized, e.g., as follows: the $\pm\,2\sigma$-intervals centered around the true para-
meter-values, as shown in Fig. 5 for measurement-noise equally distributed
over the interval -0.5 mmHg $\leqslant \eta_V \leqslant$ 0.5 mmHg, must not overlap. Already
doubling the maximum amplitudes of measurement-noise, i.e. $|\eta_V| \leqslant 1$ mmHg,
will violate this condition.

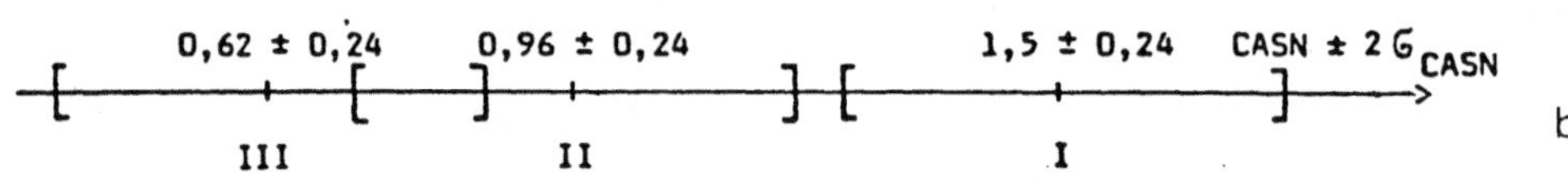

Fig. 5   ($\pm\,2\sigma$)-intervals for CASN-estimates with respect to different
degrees (I,II,III) of pathological vessels for equally distributed
mesurement-errors over the interval $\pm$0.5 mmHg (a) and $\pm$ 1.0mmHg (b)
and for a square-wave as system-excitation (200 W).

If we adjoin adjacent compliance-intervals to the CASN-values, characterizing the different degrees of illness, the following statement is true:
parameter-estimates having a distance of more than 2σ from the corresponding interval-boundaries will characterize, significantly in this sense,
one of the â priori chosen degrees of illness.

5. Conclusion

It has been shown for measurement-noise, equally distributed over $\pm$0.5mmHg,
that the accuracy of the parameter-estimates can be improved by
suitable choice of the sampling-schedule and the system-excitation, so
that the $\pm\,2\sigma$-intervals for favourably placed parameter-estimates with
respect to â priori chosen intervals of degree of illness will no more
overlap.

Because blood-pressure cannot be measured non-invasively more procise than
$\pm$ 1 mmHg to $\pm$ 2 mmHg an application of the method investigated here for
estimating the compliance-parameters needs an improvement of the accuracy
of the blood-pressure measurements at least by a factor of 2 to 4.

6. Literature

/BEKE78/ Bekey, G.A. and J.E.W. Beneken: Identification of Biological
          Systems: a Survey, Automatica 14, 1978, pp. 41-47.

/GREW76/ Grewal, M.S. and K. Glover: Identifiability of linear and Nonlinear
          Dynamical Systems. IEEE Trans. AC-21, 1976, pp. 833-837.

/MöLL81/ Möller, D.: Ein geschlossenes nichtlineares Modell zur Simu-
          lation des Kurzzeitverhaltens des Kreislaufsystems und seine
          Anwendung zur Identifikation, Springer, 1981.

/MöLL82/ Möller, D. und W.K.R. Barnikol: Ein nichtlineares mathe-
          matisches Simulationsmodell des Kardiovaskulären Systems zur
          nichtinvasiven Bestimmung der Gefäßdehnbarkeit. Funkt. Biol.
          Med. 1, 1982, pp. 183-187.

/MöLL83/ Möller, D., D. Popović and G. Thiele: Modelling, Simulation
          and Parameter-Estimation of the Human-Cardiovascular System.
          Vieweg, 1983.

/VANS83/ Vansteenkiste, G.C. and P.C. Young (Eds.): Modelling and Data
          Analysis in Biotechnology and Medical Engineering. North-
          Holland, 1983.

/TANH85/ Tanha, A., H. Maftoon: Schätzung der Parameter dynamischer
          Systeme am Beispiel der Compliance-Parameter eines nicht-
          linearen Modells des geregelten Kardiovaskulären Systems,
          Dipl.-Arbeit, Universität Bremen, 1985.

/THIE84/ Thiele, G., D. Möller, D. Popović: Probleme bei der Schätzung
          der Parameter eines nichtlinearen Modells des physiologisch
          geschlossenen Kardiovaskulären Systems. In D. Möller (Ed.),
          Proc. Systemanalyse biologischer Prozesse, 1. Ebernburger
          Gespräch, Bad Münster am Stein-Ebernburg, 1984, Med. Inf. u.
          Statistik, Bd. 52, Springer Verlag, 1984, pp. 147-157.

/MöLL85/ Möller, D., D. Popović, G. Thiele: Reliability of Parameter
          Estimation Methods applied to the Identification of Biome-
          dical Multi-Compartment Systems. In H.A. Barker, P.C. Young
          (Eds.), Prepr. 7. IFAC-Symposium Identification and System
          Parameter Estimation, vol. 2, York 1985, Pergamon Press,
          Oxford, pp. 1385-1390.

Some aspects of the application of Neurodynamical Models for the Simulation of Central Regulation and Dysregulation

Oskar Hoffmann, Gießen

Summary: A model of short-time behaviour of cardiovascular control is described representing medullary centres by two randomly connected neural nets. The network parameters are adjusted to obtain a system reaction which, under normal conditions, coincides with a reference model well known from the literature. The effect of variations of the net parameters upon steady state arterial blood pressure, heart rate and peripheral resistance is examined. The ABP level is insensible even against great deviations from the nominal values of the net representing the motor nucleus of the vagus.

Introduction

Physiological models including central controlling structures often treat the controller as a "black box" and describe the control mechanism only by transfer functions. Disturbances of central regulation, a phenomena of outstanding importance in neurosurgery /6/, are rarely manageable by these models. For an elevated intracranial pressure (ICP) central dysregulation of cadiovascular control has been simulated by assuming dislocations of the baroreceptor feedback loop characteristics /3/. A satisfactory physiological explanation of this proceeding can hardly be given. Furthermore midbrain dysfunction has been observed even at moderately elevated ICP /10/. Therefore in this study the attempt is made to combine a neural net model with a simple model of the cardiovascular system.

The Model

The model of the cardiovascular system includes the left ventricle of the heart, regarded as a rhythmically acting pump, the aorta as an elastic chamber and the peripheral resistance (PR). The cardiac output *per minute* (HMV) is expressed as a quadratic function of the heart rate (HR) /4/. The arterial blood pressure (ABP) is controlled by adjusting HR and PR. The model using the baroreceptor feedback loop characteristics developed by MÖLLER /5/ is regarded as a reference model.

In the proposed new model neural networks are used to represent
the medullary centres. These nerve nets (Fig. 2) are composed of
formal neurons /8/. A formal neuron has a large number of inputs,
taking on the values 1 (active) and -1 (inactive). Using the
synaptic weights w_i the post-synaptic potential is defined by

$$p = \sum_{i=1}^{k} w_i x_i$$

The formal neuron fires (output x=1) when the post-synaptic
potential exceeds the threshold h.

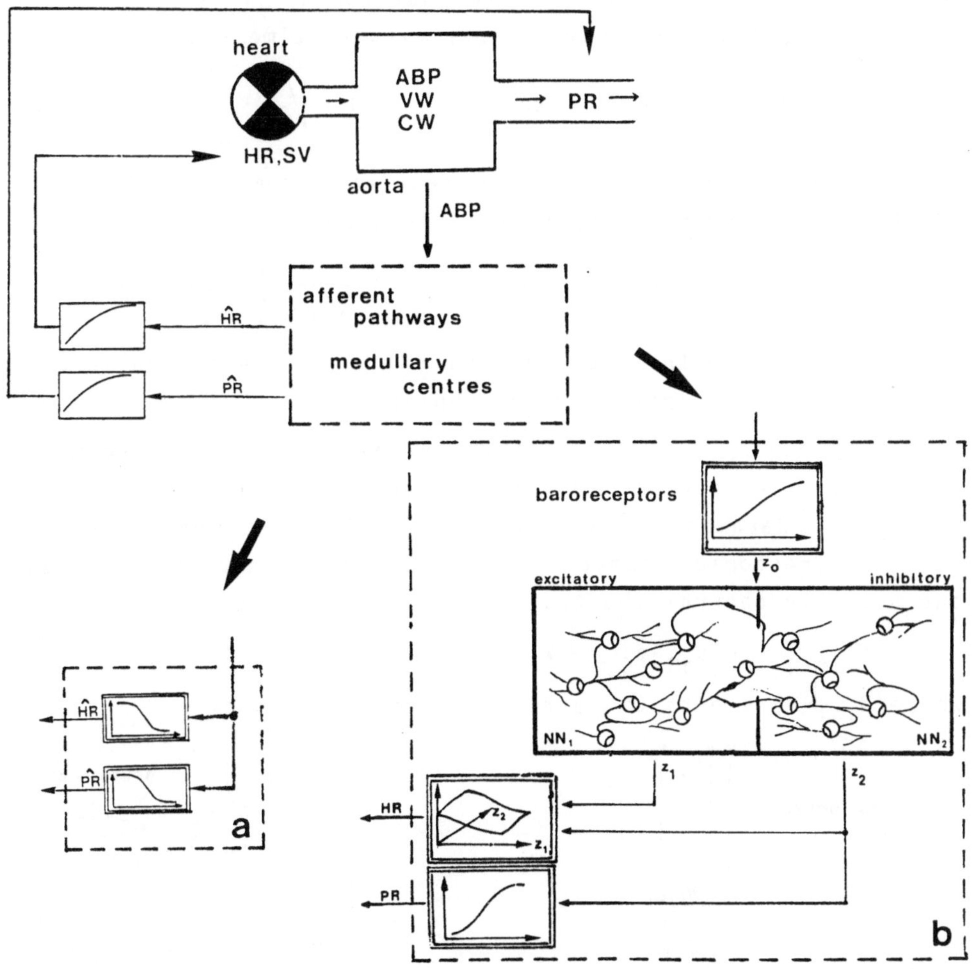

Fig. 1: Model of the cardiovascular system including central
controller structures :
a) Baroreceptor feedback loop according to MÖLLER /5/.
b) Medullary centres composed by two neural nets.

Now let us consider a net composed of n formal neurons N_i. The synaptic weight of the connection between the output of N_j and N_i is denoted by w_{ij}. The threshold of N_i is denoted by h_i. The quantity

$$z = \frac{1}{n} \sum_{i=1}^{n} x_i$$

is called the activity level of the nerve net, ranging between -1 and 1. The local interactions between the nerve cells can be regarded as random /1,2,9/. Assuming normally distributed h_i's, a sufficiently large n and all neurons working synchroneously at discrete times AMARI /1,2/ proved that the activity level at time t+1 can be calculated from the activity at time t by

$$z_{t+1} = \Phi(W \cdot z_t - H) \qquad \text{where} \qquad \Phi(x) = \sqrt{\frac{2}{\pi}} \cdot \int_0^x e^{-t^2/2} \, dt \; .$$

The macroscopic quantities W and H, resulting from the statistical parameters of the w_{ij}- and h_i- distribution are defined by

$$W = \frac{n \cdot \overline{w}}{\sqrt{n \cdot \sigma_w^2 + \sigma_h^2}} \qquad \text{and} \qquad H = \frac{\overline{h}}{\sqrt{n \cdot \sigma_w^2 + \sigma_h^2}}$$

where $\overline{w}$ and $\overline{h}$ are the averages and σ_w^2 and σ_h^2 are the variances of the synaptic weights and thresholds.

For the purpose of modelling the central controller it seems necessary to couple at least two nerve nets with different macroscopic parameters. Similar to a single neural net we can derive macroscopic parameters expressing the connections between the nets. In general the transition function is given by

$$z_{i,t+1} = \Phi(W_{i1} \cdot z_{1,t} + W_{i2} \cdot z_{2,t} + \ldots + W_{ik} \cdot z_{k,t} - H_i)$$

where W_{ij} represents the effect of net NN_j upon NN_i.

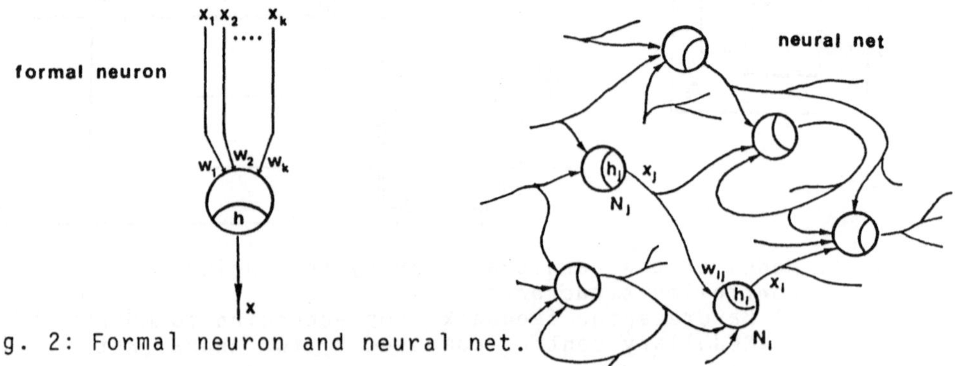

Fig. 2: Formal neuron and neural net.

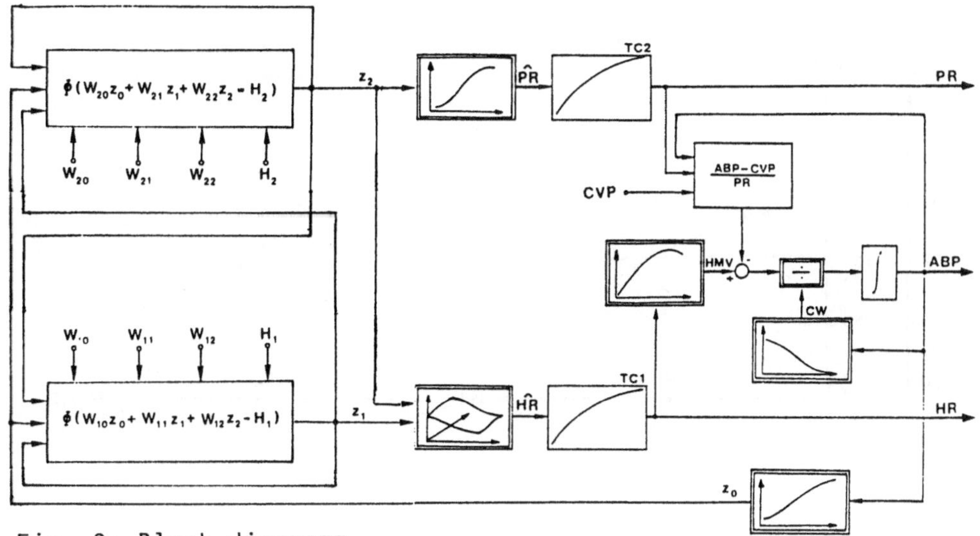

Fig. 3: Block diagramm.

The proposed model uses two nerve nets which are stimulated by
the activity z_0 resulting from the baroreceptors. The activity
levels z_1 and z_2 are transmitted via efferent pathways and are
transformed into adjustments of HR and PR.

Results

The parameters of the neural nets and the functions transforming
the neural activity into changes of HR and PR have been adjusted
to obtain the same steady state values as by the reference model.
Disturbances of the system by sudden increase or decrease of HR
or PR result in identical reactions in the reference model and
in the model proposed. The effect of changes of the neuronal
thresholds and synaptic weights upon ABP, HR and PR are demonstrated
in Fig. 4. It is obvious that the level of the controlled variable
(ABP) is insensible even against large deviations from the nominal
values of net 1, representing the motor nucleus of the vagus. In
contrast variation of parameters of net 2, repreenting "cardio-
accelerator" and "vasoconstrictor" centre, immediately changes
the ABP level.

Discussion

The simulation of a biomedical system under normal conditions is
of less interest than the adequate description of the pathological

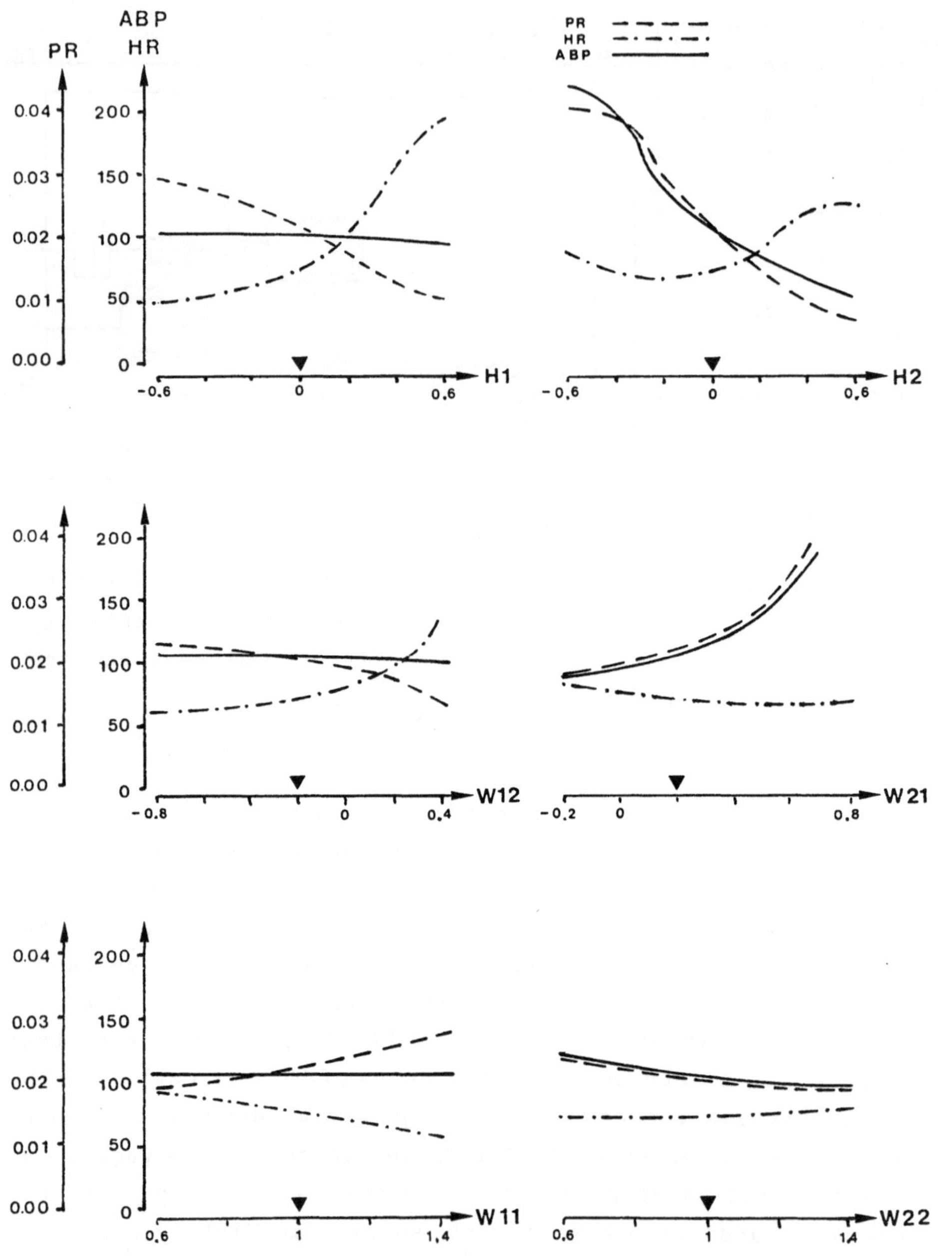

Fig. 4: Effects of variation of net parameters upon steady
state level of ABP, HR and PR. (▼: nominal values).

behaviour when neurosurgically relevant aspects are concerned. The
proposed model represents the central controller by two coupled
neural networks. It offers a better entry to questions of central
dysregulation than models treating the control centres as "black
boxes". The development of pathological behaviour can be explained
by alterations of synaptic functions and of thresholds. The
transition of regulation to dysregulation, and vice versa, demands
a plasticity of the neural nets, i.e. a reorganisation of synaptic
connections and weights and of thresholds /8/. A detailed concept
to explain this self-organisation leading to structural and
functional changes of the system is not known. The existence of
brain plasticity, even in the elderly, has been demonstrated im-
pressively by PIA /7/. Under this aspect the application of neurodynamical models for simulation of central regulation and dysregulation seems to be advantageous for studies connected with
adaption processes of the central nervous system.

References

1.   Amari, S.: A method of statistical neurodynamics. Kybernetik
     14 (1974) 201-215.
2.   Amari, S.: A mathematical theory of nerve nets. Adv. in
     Biophys. 6 (1974) 75-120.
3.   Hoffmann, O.: Ein mathematisches Modell zur Simulation und
     Analyse der intrakraniellen Liquor- und Hämodynamik. Eine
     medizinisch-theoretische Studie. Habilitationsschrift,
     Gießen 1985
4.   Kirchheim, H.: Kreislaufregulation. In: Kreislaufphysiologie
     (R. Busse, ed.), Stuttgart-New York, Thieme 1982, 167-210
5.   Möller, D.: Ein geschlossenes nichtlineares Modell zur
     Simulation des Kurzzeitverhaltens des Kreislaufsystems und
     seine Anwendung zur Identifikation. Berlin-Heidelberg-
     New York, Springer 1981.
6.   Pia, H.W.: Central dysregulation. Z. Neurol. 204 (1973) 1-21.
7.   Pia, H.W.: Plasticity of the central nervous system - a
     neurosurgeons experience of cerebral compensation and
     decompensation. Acta Neurochir. 77 (1985) 81-102
8.   Theil, S.: Neurokybernetik. In: Neurobiologie (D. Biesold,
     ed.). Stuttgart-New York, Fischer 1977, 793-843.
9.   Wilson, H.R., Cowan, J.D.: Excitatory and inhibitory inter-
     actions in localized populations of model neurons. Biophys.
     Journal 12 (1972) 1-24.
10. Zierski, J.: Intracranial pressure and brain stem involvement.
     Habilitationsschrift, Gießen 1985.

Measuring Symptoms in Parkinson's Disease with a tracking device

Selim S. Hacisalihzade, Carlo Albani, Markus A. Mueller, Zurich

Summary: A simple computer based visual tracking device is presented which allows to measure different motor performances in Parkinsonian patients. The device uses a computer to drive the reference signal which is tracked by the thumb movements of the patient. The computer provides the signals for different tracking conditions, records the tracking performance and analyses the recorded data. Three examples are used to show how the device can be used to measure different Parkinsonian symptoms.

Zuammenfassung: Ein einfaches rechnergesteuertes Geraet fuer visuelle Verfolgungsaufgaben wird vorgestellt. Es erlaubt die Messung verschiedener motorischer Faehigkeiten von Parkinson-Patienten. Ein Rechner steuert ein Referenzsignal, das durch Daumenbewegungen des Patienten verfolgt werden soll. Weiter kontrolliert der Rechner Stoerungen, die die Verfolgung erschweren, registriert die Daumenbewegungen des Patienten und analysiert die so aufgenommenen Daten. Drei Beispiele zeigen die Messung einiger Symptome der Parkinsonschen Krankheit mit Hilfe des Geraetes.

1. Introduction

In clinical practice the symptoms of Parkinsonian patients are assessed by different rating scales [1], [2]. However, these evaluations are only qualitative and depend heavily on the subjective rating of the examiners. Therefore, there has been a long quest to find objective ways of measuring motor performances and neurological functions [3], [4], [5].

Many measuring setups have been proposed, most of them relying on electromechanical configurations. Since most of them were developed in the beginnings of the sixties, they did not make use of the computer facilities available today. Tracking tasks to assess Parkinsonian deficits have been used as early as 1962 [6] and newer versions can be found in the literature [7].

Since motor tasks can be performed using different motor strategies, measurements using movements with several degrees of freedom are usually difficult to analyse. The measurements proposed in this paper use the displacement of the end phalanx of the thumb to perform the tracking tasks. This movement has a single degree of freedom and the peripheral control mechanisms have been analysed in detail [8].

A slightly modified version of this paper has been presented at Medinfo 86

"

2. The Method and the Device

The thumb phalanx is attached to the shaft of a direct current motor. The angular position of the shaft is measured by a potentiometer driven by the same shaft. The hand and the forearm are secured to ensure that the tracking is performed by the flexion and extension movements of the end phalanx of the thumb alone (see Fig. 1).

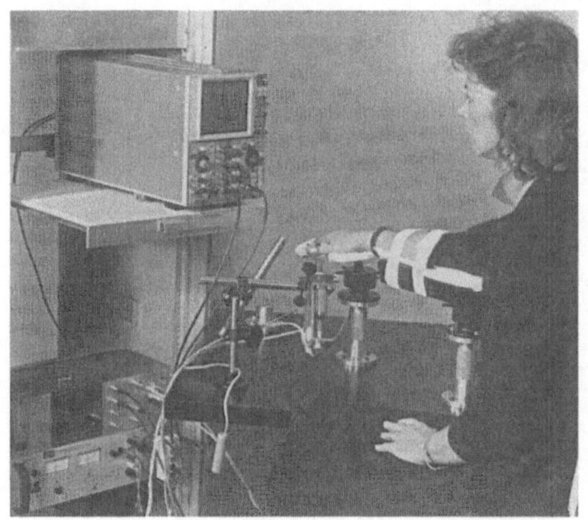

Fig. 1

The reference input of the tracking task is a small target which moves on the vertical axis of an oscilloscope screen. The task is to move a point of light up and down on the screen so that it stays in the middle of the target. The position of the point of light corresponds to the angular position of the shaft which can be rotated by the movements of the end phalanx of the thumb.

The motor can produce disturbances of the thumb position by applying a variable torque. It is possible to perform measurements with this configuration with or without disturbing torque as well as measurements in which the actual position of the thumb is visible only for a fraction of the time (visual or somatostatical open loop conditions.

The computer is used to control the movements of the target as well as the current which produces the torque of the motor. It also records the tracking performance of the subject which is analysed at the end of the measurement. See Fig. 2 for a functional diagram of the measurement layout. This setup can be used to get quantitative measures of various symptoms.

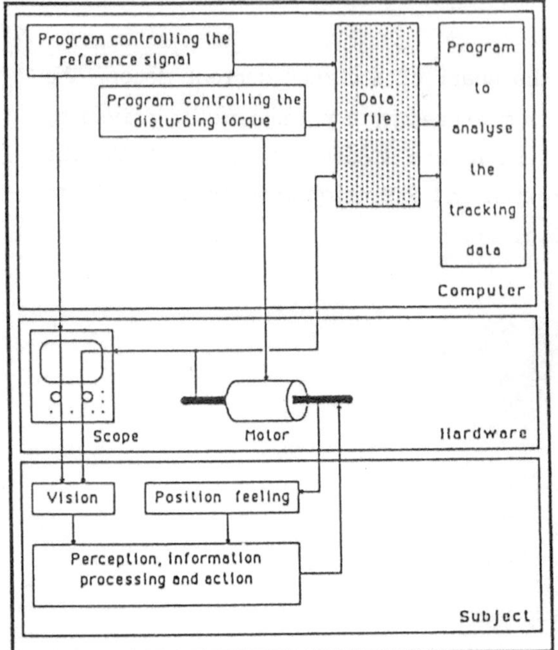

Fig. 2

3. Examples

In a simple measurement without any disturbing torque from the motor, the subject has to track the step motions of the reference signal. The steps are 6 cm high on the oscilloscope screen which corresponds to an angle of about 7° from 50 cm distance and continue for four to six seconds (equally distributed) with again four to six seconds between two steps. A single session goes on for about forty seconds and seven such sessions with a minute long intervals between each constitute a measurement. The tracking data is sampled a hundred times a second.

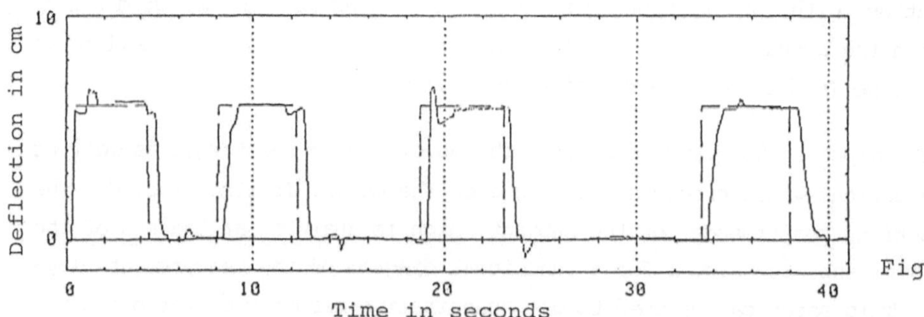

Fig. 3

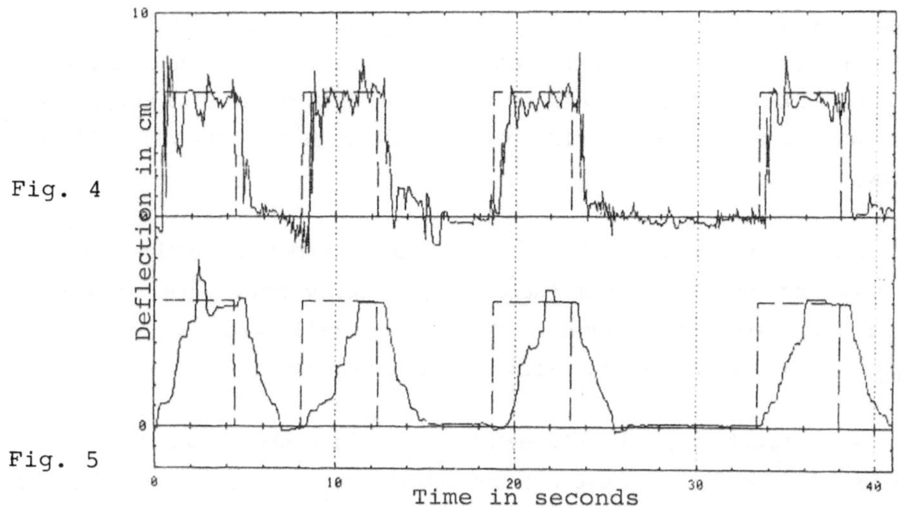

Fig. 4

Fig. 5

Fig. 3 depicts the outcome of such a session with the right hand of a control subject. The broken line indicates the time history of the reference signal and the continuous line the response of the subject. Fig. 4 shows the outcome of an identical session with a Parkinsonian patient presenting a tremor of the upper extremities as the main symptom. As can be seen, he suffers from excessive tremor of the right thumb. Fig. 5 shows the performance of another patient with a pronounced slowing down of motor activity in the right hand in an identical session as in Fig. 3 and 4.

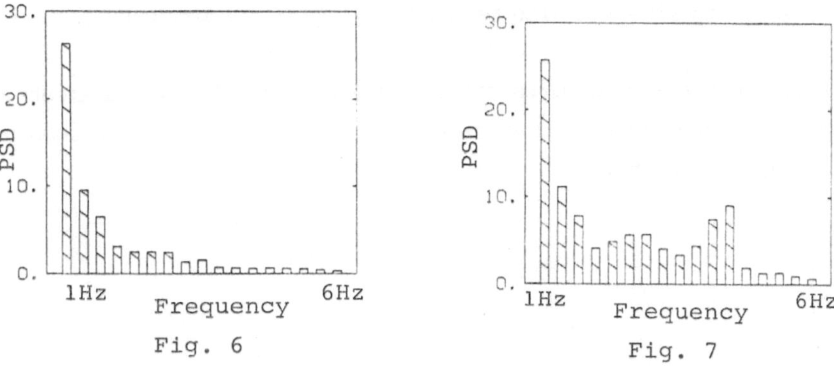

Fig. 6

Fig. 7

In order to quantify, compare and classify the tremor one has to consider the power spectral densities (PSD) of the time histories depicted in Fig. 3 and 4. The PSD of the control subject's session is shown in Fig. 6, whereas Fig. 7 depicts the PSD of the Parkinsonian patient. The

tremor manifests itself as higher frequency components (especially at 5
Hz) present in the PSD of the patient. One can now, for instance, define
a tremor index as the integral of the PSD with given lower and upper limits which corresponds to the energy of the signal in that specific frequency range.

The time history of the tremor patient's session defies further analysis
in this form just because of these high frequency components. In order to
get rid of them, the response is filtered digitally with a tenth order
tapped delay low pass filter with a limit frequency of 1 Hz.

Two very important parameters among others which can be derived from the
measurements described above are the lengths of the initiation time and
movement time. The initiation time is defined as the time the thumb of
the subject needs to move from zero to 5% of the step size. Analogously,
the movement time is defined as the time the thumb needs to move from
5% to 95% of the step size. A related parameter is the speed with which
the movements are executed. The initiation time (IT), the movement time
(MT) and the speed together give a very good idea about the slowing down
of motor action in Parkinson's disease (hypokinesia) due to rigor and
bradykinesia.

Another parameter which can be derived from these measurements is the area between the time history of the reference signal and the response of
the subject. Its size can also be used to quantify bradykinesia.

Fig. 8 synopsizes and compares selected results of the measurements on
the three above mentioned subjects. The shown values are averages over
seven similar sessions.

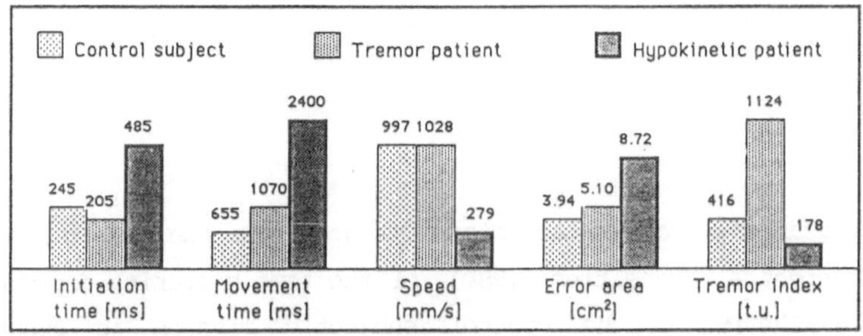

Fig. 8

A quick glance at Fig. 8 shows that

- the initiation times of the control subject and the tremor patient are similar and that of the hypokinetic patient is about two times as long

- the movement time of the tremor patient is about 60% longer, that of the hypokinetic patient three to four times as long as the movement time of the control subject

- a similar, but less pronounced relation holds between the error areas of the respective subjects

- the speed with which the control subject and the tremor patient execute the movements are comparable, whereas the hypokinetic patient moves about three times slower.

- the tremor index of the tremor patient is about three times higher than that of the control subject. Whereas, the tremor index of the hypokinetic patient is about half as much (high frequency components are absent due to the slowing down of motor activity).

4. State of the Project and Further Research

The described setup was designed to measure the different symptoms of Parkinsonian patients and the effect of the drug therapies in an objective way. The symptoms of Parkinson's disease can be relieved by dopaminergic agents. The dosage of the used drugs has long been the issue of a controversial discussion . Therapy with low dosage regimens results in periods where the therapeutic effect is lost ("off state"). High dosage regimens result in immediate side effects like dyskinesia as well as cumulative long term ones. Therefore, one is inclined to prescribe therapies with an individual dosage scheme [9]. In order for the therapies to be optimal,the pharmacodynamic response of the individual must be measured. This can be accomplished with the above presented device. A mathematical model of the effect of levodopa on the symptoms of the individual patient can be built with the aid of the mentioned measurements. The model can in turn be used to optimize the dosage scheme individually.

<u>Acknowledgements</u>

This work was supported in part by the Swiss NSF grant 3.889.0.85. The authors would also like to thank H. Davatz and P. Lansky for their help in evaluation of the measurements.

<u>References</u>

[1]   D.D. Webster, Modern Treatment, Vol. 5, pp. 257-282, 1968.

[2]   R.C. Duvoisin, "Monoamines Noyaux Gris Centraux et Syndrome de Parkinson", J. Ajuriaguerra ed., pp. 313-325, Mosson, Paris, 1971.

[3]   W.W. Tourtelotte, A.F. Haerer, J.F. Simpson, J.W. Kuzma, J. Sikorski, "Quantitative Clinical Neurological Testing", New York Acad. Sci., Vol. 122, pp. 480-505, 1965.

[4]   A.R. Potvin, W.W. Tourtelotte, J.S. Daily, J.E. Walker, J.W. Albers, W.G. Henderson, D.N. Snyder, " Simulated Activities of Daily Living Examination", Arch.Phys. Med. Rehab., Vol.53, pp. 476-487, 1972.

[5]   J.W. Albers, A.R. Potvin, W.W. Tourtelotte, R.W. Pew, R.S. Stribley, "Quantification of Hand Tremor in the Clinical Neurological Examination", IEEE Trans. Biomed. Eng., Vol. BME-20, pp. 27-37, 1973.

[6]   W.D.A. Bowen, M.M. Hoehn, M.D. Yahr, "Cerebral Dominance in Relation to Tracking and Tapping Performance in Patients with Parkinsonism", Neurology, Minneapolis, Vol. 22, pp. 32-39, 1962.

[7]   K.A. Flowers, "Visual 'Closed-Loop' and 'Open-Loop' Characteristics of Voluntary Movement in Patients with Parkinsonism and Intention Tremor", Brain, Vol. 99, pp. 269-310, 1976.

[8]   C.D. Marsden, P.A. Merton, H.B. Morton, "Servo Action in the Human Thumb", J. Physiol., Vol. 257, pp. 1-41, 1976.

[9]   C. Albani, R. Asper, S.S. Hacisalihzade, G. Baumgartner, "Individual Levodopa Therapy in Parkinson's Disease", Advances in Neurology: Proceedings of the 8th International Parkinson's Disease, New York, 1985), Raven Press.

Round Table Discussion

Basic Questions of Modelling

Gerhard K. Wolf, Heidelberg

This discussion is dedicated to the fundamental problems of
scientific reasoning as usually dealt with by philosophy of
science. From the experts point of view on modelling biological
and medical systems, it was tried to find out which parts of
these philosophies and of additional principles should be used
in our context.

The usually mentioned philosophical views on scientific
reasoning are known as
- deductivism,
- inductivism,
- falsificationism,
- falsificationism with additional confirmation
and finally the
- theory of paradigmatic revolutions.

<u>Deductive Reasoning</u>

Deduction is the classical way of reasoning which goes back to
the old greek philosophers. But even now, for many purposes it
is the only way used.

As SCHÖNE pointed out during discussion, very often a merely
deductive method to construct models is used: the theories of
physics are used to derive the model. From the model then, the
expected behaviour of a real system is explained.

The most simple logical description of such a conclusion is the
socalled 'modus ponendo ponens': $A \to B$, $A \frown B$, in words: If A
holds, than B, now A is given, therefore B is concluded. A
classical realization of this kind of deduction is the modus
Barbara, which goes back to Aristotle: All men die, Sokrates is
a man, thus he will die.

Inductive Reasoning

Deduction yet can never synthesize new explanations for new
phenomena. Therefore deduction is not the only way of reasoning
in the area of modelling. Some participants of this discussion
feeled that in a field of medical modelling deduction usually
cannot be used because of the lack of accepted theories.

Therefore an inversion of the deduction seems to be necessary in
order to find new scientific explanations. This is called the
induction principle: $A \rightarrow B$, $B \frown A$. But this is never valuable.
Usually there exists an infinity of other explanations or models
or theories which also could lead to the result B. Therefore
inductivism as an explanation for the finding of good models or
theories must not be used.

Falsification as a Method

Falsificationism uses the fact that always holds: $A \rightarrow B$,
not B $\frown$ not A. That means a model cannot be supported by
experience, but it could be rejected if new observations
contradict a conclusion from A.

As is pointed out very often and even in this workshop (e.g. by
REICHL), there exists no observation without any prior knowledge
and some theory of observation. We have to make measurable what
was not. One example for this work was the contribution of
HACISALIHZADE in this workshop. Thus the process of verification
of models usually has different strata.

Reasonings Using Falsification And Confirmation

MÖLLER gave a description of his view to the process as a circle
which goes from the observation of the real system to the
construction of an abstract model and then to a real model. The
behaviour of which can be compared to the real system giving

eventually new and unexpected observations. Then this circle has
to be started again.

This introduces the option of the corrobation of models. Within
philosophy of science, especially POPPER gave a theory which
underlines the necessity of the process of falsification and
corrobation (confirmation).

Example to Illustrate the Philosophic Techniques Applied to
Modelling

An own simple model for the circulation of lymphocytes from
blood to the tissue and again via vessels like the thoracic duct
to the blood was used to illustrate these processes.

A differential equation describes the more or less known facts:

$$z'_t + (m+p)z_t - n = 0,$$

where z_t means the number of lymphocytes within the animal (the
system under consideration), m, p and n are the parameters of
the model whith m: rate of eliminated lymphocytes, p: the
probability to pass the thoracic duct and not another vessel and
n: the number of new lymphocytes. Then, from data of an
experiment, where the lymphocytes from the thoracic duct were
eliminated, the parameters of the model could be estimated.
$z_{t=0}$ and p are approximately known from other observations and
therefore the new estimators found by approximating the model
could be compared and were found to be satisfying. (m and n were
considered to be useful for future observations). Thus, the
model is not falsificated. But this is not a real trial to
falsificate the model. This should only be done with a new
derivation from the model, with a new experiment on a real world
a more complex experimental arrangement with observation of
lymphocytes within the thoracic duct but retransfusion of them.
This was modelled and the results were compared to the

observations. Because the model was not falsificated, it can be considered as confirmed.

We think it should be one of the duties of the constructor of a model to find experiments which are apt to falsificate this model. Another example where this was done is the contribution to this workshop given by PABST, who even gave a hint to a crucial experiment to decide between a class of models like his and the very common model for the aging of tissues.

A Call to Change Paradigms

Our example of a model makes use of differential equations. This is the most prevalent way and a good example for a paradigm at least during these first two Ebernburg workshops. Differential equations yet are only feasable if the empiric relative shows structures apt to be measured by proportional scales. The axioms of real numbers must have homomorphous counterparts in the real system. If this is not the case other methods for modelling reality have to be used. For example, grammatical representations like the Petri nets, as discussed by FUSS during this workshop or even syntax diagrams as used for our blood bank system may prove to be fruitful. RICHTER called for better suited maths in modelling biological systems. This would lead to some kind of revolution and to the replacement of the paradigm of the superiority of differential equations by a systematic view on a pluralistic modelling.

Summary and Discussion

Summary and discussion

The broad range of topics covered by the second Ebernburg conference on biological system analysis necessitates a review which outlines recent trends
and developments reflected in the contributions of the participants of this
meeting.

In the first place it is important to discuss how biological processes can
be translated into mathematical models. The mathematical tools which are
available mostly originate from physical or technical system analysis. Thus,
continuous processes are mapped on systems of differential equations, discrete
processes are described by difference equations or are simulated by use of
random number generators. In contrast to technical systems biological systems
tend to be arbitrarily complex: it is impossible, to comprise all components
of an ecosystem into a model, and even at the level of biochemistry it is not
feasible to write down all relevant elementary chemical reactions of an
enzymatic network. In addition, most technical systems operate in the neighborhood of an operating point, where the system response may be modeled by linear
functions, whereas in most biological systems nonlinear responses are essential.
Nonlinearities generate complex dynamic patterns such as limit cycles or even
chaos. These challenging dynamic properties can be analysed by mathematical
methods only if the dimension of the model system is small. Large systems can
only be explored by simulation, the term simulation referring to the numerical
solution of the differential equations. However, no general conclusions on the
dynamic behavior can be drawn based solely on simulation, since the scope of
simulation experiments is always confined to the range of parameters used.
Therefore, two objectives of biological models can be distinguished:
detection of mathematical structures underlying basic biological processes and
simulation of large systems aiming at applications in the area of biomedical
control.

The paper given by an der Heiden on the strange dynamic patterns produced by
a simple differential delay equation and by Haken and Wunderlin on synergetics
are typical examples of explanatory models. An der Heiden analyses mathematical
structures arising from a simple and often used type of a biological model
equation. This paper introduces terms of modern dynamic theory such as limit

cycles, invariant measures and chaos. Wunderlin gives an outline of the
physical theory of synergism explaining the generation of stable macroscopic
structures in systems which comprise a large number of subsystems on a
microscopic scale. Although the main applications of this theory lie in the
realm of physics, it will surely have its impact on theoretical biology,
which is still lacking a general theory at an abstract level.
A further explanatory model is given by Pabst who developes a new concept
of sleeping stem cells.

At the other extreme stands the complex model of Lieth which simulates the
dynamics of the whole biosphere. It is used to analyse the impact of fossil
fuel combustion on the fate of $CO2$ in the biosphere and thus on future global
climatic changes. The statistical problems arising when modeling multivariate
response functions, which are the building blocks of larger ecosystem models,
are treated in.the paper of Klein.

But one need not to simulate the whole world to get large models: the model
systems do not tend to become less complex if one goes down to lower
organisational levels.
Bossel presents a dynamic simulation model for the growth of trees under the
influence of pollutants. Although the mechanisms of the effect of possible
pollutants on the plant biochemistry have not been identified as yet, the
model of Bossel allows the simulation of the collapse of a plant under generel
assumptions concerning the inhibition of photosynthesis and the inhibition of
uptake of nutrients by the roots. The growth curves produced by this model
describe well the sad fate of central european forest trees.

The simulation models at the level of physiology have been developed for
practical medical purposes: Sikora, Möller, Pohl and Hennig have devised a
model of the human blood circulatory system as an aid for the development of
artificial hearts. At this point the necessity of parameteridentification and
model validation becomes quite clear. Therefore, the same authors developed
methods for the parameteridentification in cardiovascular models by model
reduction. In the paper of Tanha, Maftoon, Thiele and Möller it is shwon,
how improvements of the estimates of compliance parameters are achieved by an
appropriate experimental design.
The problems arising when estimating parameters in large dynamical systems
given as differential equations are treated in the mathematically advanced
paper of Schloeder and Bock. The numerical procedures which are developed by

the authors are based on a multiple shooting method and are apt for large
scale identification problems especially if implicit models are involved.
The performance of this method is demonstrated for the well known nonlinear
chemical reaction system of the oscillating Belousov-Jabotinsky reaction.

Whereas in the foregoing models the cardiovascular system is treated
separately, Hoffmann treats both a neurodynamical system and an cardiovascular
system in one model analysing mechanisms of central regulation and dysregulation.

Two papers are concerned with processes at the level of biochemstry. Hafner,
Berger and Borchard model membrane transport processes in electrophysiological
pharmacology. The authors are a team consisting of pharmacologists and a biomathematician. Experiments and model development are closely related, a
situation which is seldom met. Under these conditions the model is advanced
to a stage, where specific interactions of antiarrhythmic agents with ion
chanels can be identified by estimation of specific parameters from experimental
curves.

The paper of Reich discusses a new approach for the simulation of large
enzymatic networks by use of artificial intelligence.

Thus far, only continuous systems and processes are regarded, which are
described in terms of differential equations. When discrete systems are
simulated, one has to resort to Monte Carlo methods, i. e. generating possible
chains of events using random numbers. Gabriel uses simulations of this kind
to test a new fitness concept. Fitsch and Kaiser generate life histories of
daphnia magna.
Wosniok considers stochastic growth models and methods for estimation individual
growth curves from aggregate data.

Two papers are concerned with more general aspects of model building. Schöne
shows how to use bond diagrams for the description of open thermodynmic processes.
Vogelsänger presents a mathematical method of modeling biological structure
and control systems. Structural aspects which are most important in biological
systems are most difficult to incorporate into model systems and are omitted
in most models. Therefore, the approach given by Vogelsang is an important
step towards more realistic simulation.

When analysing biological systems, the first steps are the acquisition of
data and their statistical analysis. Two papers are devoted to this subject.
Hacisalihzade, Albani and Mueller report on their experience with a new
tracking device when measuring symptons in Parkinson's disease. Ranft
introduces methods of texture analysis which are applied to the interpretation of ultrasonic images of the liver.

A biomathematician should always look for new mathematical concepts arising
in the hope to be rewarded by a new approach to biological system analysis.
To this end, an invited lecture was given by Fuss on Petri nets which are
widely used for the design of large networks. Although the concepts given
are not easily accepted by those more familiar with continuous system dynamics
on the basic of differential equations, petri nets may be applied to the
modeling of organization of an organism in space and time.

List of Speakers

Bossel, H. Prof. Dr.	FB 17, FG Umweltsystemanalyse, Gesamthoch- schule Kassel, Mönckebergstr. 21, D-3500 Kassel
Fitsch, V. Dipl.-Biol.	Lehrstuhl für Biologie (Ökologie) der RWTH-Aachen, Kopernihusstr. 16, D-5100 Aachen
Fuss, H. Dr.	Institut für Methodische Grundlagen (F 1) GMD Bonn, Schloss Birlinghoven, D-5205 Sankt Augustin 1
Gabriel, W. Dr.	Max-Planck-Institut für Limnologie, Postfach 165, D-2320 Plön
Hacisalihzade, S.S. Dipl.-Ing.	Institut für Automation, ETH-Zürich, Physikstr. 3, CH-8092 Zürich
Hafner, D. Dr.	Institut für Pharmakologie, Universität Düsseldorf, Moorenstr. 5, D-4000 Düsseldorf
an der Heiden, U. PD Dr.	Fachbereich Biologie, Universität Bremen, Achterstr., D-2800 Bremen 33
Hoffmann, O. PD Dr.	Neurochirurgische Univ.-Klinik, Justus- Liebig-Universität Gießen, Klinikstr. 29, D-6300 Gießen
Klein, G. Dr.	Fachbereich Biologie/Chemie, Universität Osnabrück, Barbarastr. 11, D-4500 Osnabrück
Lieth, H. Prof. Dr.	Fachbereich Biologie/Chemie, Universität Osnabrück, Barbarastr. 11, D-4500 Osnabrück
Pabst, G. Dr.	Abteilung für Klinische Physiologie und Arbeitsmedizin, Universität Ulm, Oberer Eselsberg, D-7900 Ulm
Pohl, V. Dipl.-Ing.	FB Elektrotechnik, Universität Bremen, Achterstr., D-2800 Bremen 33
Ranft, U. Dr.	Institut für Biometrie, Medizinische Hoch- schule Hannover, Konstanty-Gutschow-Str. 8, D-3000 Hannover 61
Reichl, J. Prof. Dr.	Institut für Tierernährung, Fachgebiet: Ernährungsphysiologie, Universität Hohenheim, Emil-Wolff-Str. 10, D-7000 Stuttgart 70

Schlöder, J.P. Dr.

Institut für Angewandte Mathematik,
SFB 72, Universität Bonn, Wegelerstr. 10,
D-5300 Bonn 1

Schöne, A. Prof. Dr.

FB Produktionstechnik, Universität Bremen,
Achterstr., D-2800 Bremen

Sikora, T. Dipl.-Ing.

Chirurgische Klinik und Poliklinik,
Klinikum Charlottenburg, Spandauer
Damm 130, D-1000 Berlin 19

Tanha, A.R. Dipl.-Ing.

FB Elektrotechnik, Universität Bremen,
Achterstr., D-2800 Bremen 33

Vogelsaenger, T. Dr.

FB Elektrotechnik, Universität Siegen,
Hölderlinstr. 3, D-5900 Siegen 21

Wolf, G.K. PD Dr.

Institut für Medizinische Dokumentation,
Statistik und Datenverarbeitung,
Universität Heidelberg, Im Neuenheimer
Feld 325, D-6900 Heidelberg

Wosniok, W. Dipl.-Math.

FB Mathematik und Informatik, Universität
Bremen, Achterstr., D-2800 Bremen 33

Wunderlin, A. PD Dr.

Institut für Theoretische Physik,
Universität Stuttgart, Pfaffenwaldring 57,
D-7000 Stuttgart 80

Index Register

Tracking Device   188,189,190,
                   204
Transpiration   47,49,56
Tree   46,47,54,202
Tremor-   191,192
  index   193
Trial and Error   66